THE HORMONES
IN
HUMAN
REPRODUCTION

GEORGE W. CORNER

PRINCETON
PRINCETON UNIVERSITY PRESS
London: Geoffrey Cumberlege, Oxford University Press
1947

Revised edition, 1947
Fourth printing

Copyright, 1942, by Princeton University Press
Printed in the United States of America

· ·

Publication of this book has been aided by
the Louis Clark Vanuxem Foundation

✌§ PREFACE §✍

THIS book represents, with considerable additions, the substance of the Vanuxem Lectures, given at Princeton University in February 1942. The invitation to be Vanuxem Lecturer carried with it the expressed wish of the Committee that I should discuss the hormones of the reproductive system for the benefit of a general audience, assuming on the part of my hearers no familiarity with biology. This imposed no easy task, for it called upon me to describe some of the most intricate and elaborate mechanisms of the body, to listeners who perhaps had never seen the organs and tissues in which these activities take place. The structure of the living cells and the manner in which they are put together to form the organs are matters not merely so unfamiliar, but actually even so daunting to most people, as to create serious difficulties for the biologist and physician who tries to explain his work. For the first time in my life I could have wished I were an astronomer or physicist, for the heavenly spheres, their orbits and attractions, and even such matters as warps in space and corpuscles of light can be described to a certain extent in terms of the workshop and the household; but how can we explain the marvels of the human egg or the action of an estrogenic hormone without a background of cellular biology? My only recourse has been to begin at the very beginning, to devote as many as three chapters to general preparation for actual discussion of the hormones, and at every step to explain and illustrate the underlying anatomy and physiology as clearly as possible.

This is, to the best of my knowledge, the first time an American university has devoted one of the great endowed lectureships to the subject of human reproduction. A few years ago it might even have been impossible to break through the old conventions that hampered free public discussion of this subject. We have a tradition that sex and reproduction must be attended by privacy, dignity and romance. It is a good tradition, provided we add a fourth attribute, namely understanding; for otherwise the fundamental life activities concerned in sex may become involved in fears, inhibitions and blind taboos. I emphasize the importance—nay even the necessity—of instruction and understanding in matters of sex, in case there are still among my readers some who are troubled by our free discussion of intimate functions, and especially in case it seems to them that the dignity and the romance of life are threatened by frank acceptance of the animal nature of mankind or by our use of other creatures to explain human affairs. There is of course no denying that man is an animal, and since human physiology cannot always be subjected to direct experiment (particularly in this field of investigation), we must study the lower animals not only for their own intrinsic interest but also in order to understand ourselves. It is equally true that man is more than an animal. The ape, the tiger, and the worm mate and reproduce their kind, and so do human beings, but only man tries to understand what he is doing and why he does it. In such understanding and in right living based upon knowledge lies our best hope of attaining dignity, honor and beauty in the physical life of mankind.

A book of this kind rests upon the laborious work of many scientific investigators. The author, in drawing freely upon the writings of his colleagues, has endeavored to acknowledge their contributions as fully as possible, by mention in the text, footnotes and legends. References however are necessarily limited; readers who wish to consult the original literature will find full bibliographies in Appendix II, note 1.

Many fellow workers who have generously permitted the use of illustrations, as indicated in text and legends, deserve especial thanks.

PREFACE

The quotation at the head of Chapter I is from *Two Lives*, by William Ellery Leonard, copyright 1922, 1925, by permission of the Viking Press, Inc., New York. The quotation from C. Day Lewis's translation of Virgil's *Georgics*, in a footnote to Chapter III, is used by permission of Jonathan Cape, Limited, London and Toronto.

The author's wife, Betsy Copping Corner, and his son, Dr. George W. Corner, Jr., have given unfailing encouragement and have been patient and thoughtful critics. Mr. Arthur G. Rever has been good enough to read the manuscript and has made useful suggestions.

The author's researches upon the menstrual cycle of monkeys, cited in this book, were aided by grants to the University of Rochester by the Rockefeller Foundation and the John and Mary R. Markle Foundation.

GEORGE W. CORNER

Carnegie Institution of Washington
Department of Embryology, Baltimore

⚜ CONTENTS ⚜

Simple division into parts a frequent mode of reproduction in lower animals; necessity of egg and sperm cells in higher and more complicated creatures; the participation of two individuals, male and female, essential to the process in all higher animals; in mammals, including mankind, the fertilized egg sheltered and nourished within the mother's body; correlation of the various organs of the reproductive system to this end by action of chemical substances (hormones) made in the sex glands.

The egg a cell growing in a cavity (follicle) in the ovary; its progress, after discharge from the ovary, via the oviduct to the uterus; its implantation in the uterus, if fertilized by a sperm cell; division into many cells and development into an embryo; nourishment from the mother's blood during growth in the uterus, through an organ of attachment, the placenta.

CONTENTS

interaction between the ovaries and the pituitary gland. Menstruation a periodic breakdown of the uterine lining (endometrium) when the corpus luteum retrogresses. Occurrence, however, of anovulatory cycles, without a corpus luteum, and without "premenstrual" changes. Explanation of the bleeding as due to shutting off of the coiled arteries of the endometrium caused by deprivation of estrogenic hormone or of progesterone; bleeding due to progesterone deprivation believed to be a special case of estrin-deprivation bleeding. Theory of the menstrual cycle based on these ideas. The significance of menstruation unknown.

Calculation of the quantities of the two hormones produced in the ovaries and the rate at which they are secreted; in the case of the corpus luteum, discussion of such questions as the amount of hormone made by a single cell, the amount made by the whole gland in one day, and divers other matters of interest concerning the quantitative aspect of ovarian function.

The maintenance of pregnancy a complex affair, dependent partly on the hormones. The placenta as a source of gonadotrophic and estrogenic hormones; progesterone also apparently made by the human placenta. Lactation induced by a special hormone of the pituitary gland.

The testis constructed of tubules in which the sperm cells are made; the interstitial cells. The seminal ducts, seminal vesicles, and prostate gland under control of the testis through its hormone. Secondary sex characters described and shown to be controlled by the testis. Chemistry and effects of the androgenic hormones.

❧ LIST OF PLATES ❧

LIST OF TEXT FIGURES

LIST OF TEXT FIGURES

*THE PLACE OF THE HIGHER ANIMALS, & OF
MANKIND IN PARTICULAR, IN THE GENERAL
SCHEME OF ANIMAL REPRODUCTION*

> "of the cell, the wondrous seed
> Becoming plant and animal and mind
> Unerringly forever after its kind,
> In its omnipotence, in flower and weed
> And beast and bird and fish, and many a breed
> Of man and woman, from all years behind
> Building its future."

WILLIAM ELLERY LEONARD, *Two Lives.*

CHAPTER I

THE PLACE OF THE HIGHER ANIMALS, & OF MANKIND IN PARTICULAR, IN THE GENERAL SCHEME OF ANIMAL REPRODUCTION

AMONG the life that swarms in our southern waters, there is a charming tiny animal called Cothurnia, the buskin animalcule. These creatures cling by thousands to the vegetation on wharf piles in our harbors, and can be brought into the laboratory on a bit of seaweed in a drop of water. Because a single Cothurnia is much smaller than the printed period at the end of this sentence, it must be watched through the microscope (Fig. 1). It consists of a graceful transparent cup (formed more like a wineglass than the classical buskin from which it got its name) which is attached by its stem to some larger object. Inside the cup and fixed to its base is a single animal cell, shaped like a trumpet. While the stem sways gently in the water, the cell projects from the cup. Into its open gullet particles of food are swept by a brush of beating lashes or cilia and drift down into the jelly-like cell substance until they are dissolved and digested.

This simple career of food-gathering is interrupted from time to time by a few hours devoted to reproduction. Our pretty little trumpet withdraws itself inside the cup, rounds up a bit, and slowly separates into two cells by dividing lengthwise. For a time, both cells resume the task of feeding, but afterward one of them retires into the cup and begins a struggle to get away. It pulls so strongly, indeed, upon its stalk that its shape changes; from a trumpet it becomes a shoe. The cilia change position so that they can serve for propulsion in swimming. At last the cell breaks from its attachment and slips out into the sea, ultimately to settle

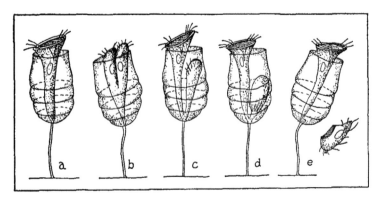

Fɪɢ. 1. The one-celled animal *Cothurnia,* reproducing itself by simple division. The parent animal is seen at *a.* From *b* to *d,* successive stages of division. In *e* the daughter cell has freed itself and is swimming away, to settle in a new location. Greatly magnified.

down upon a near-by strand of seaweed, or perhaps (venturing greatly) as far away as the next timber of the wharf.

All this makes no difference to the first cell; it undergoes no pregnancy, feels no pangs while giving birth and takes no responsibility to nurse, guard, or educate its offspring. The latter in turn asks nothing at all of its parent, and never realizes the disadvantages of birth at so low a level of organization, one of which is that the newborn cell faces immediately and alone all the dangers of its world. The infant mortality of Cothurnia must be enormous, for there are many enemies and risks, but what of that? The parent can easily split off another cell, and in spite of the wastage it is more economical (if all you want is a one-celled child) to breed by excess production than by the intricate process through which man and the higher animals turn out their limited output of complex and troublesome offspring.

Other unicellular animals have developed variations of the process of reproduction by division. Sometimes they do not divide into two equal cells, but put out their daughter cells as mere buds which break off while small and only later reach "adult" size (Fig. 2). Sometimes the parent animal breaks up by multiple fission into a relatively large number of very small daughter cells resembling spores (Fig. 3).

Reproduction by fission is so easy that in the course of evolution the animal kingdom held on to it for a long time, and many animals higher than the unicellular animals made use of it. Some of the worms, for example, split in two transversely by forming an extra head from some of the segments near the middle of the body (Fig. 4). This head, with the rest of the worm that lies behind it, drops off and wriggles away, while the original worm forms a new tail at its truncated posterior end. Sometimes the worm breaks up into a whole chain of segments each of which becomes a new worm.

Reproduction by budding also continued in higher ani-

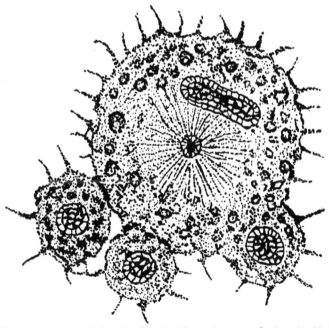

FIG. 2. A one-celled animal, *Acanthocystis*, reproducing itself by budding. 3 buds are seen. Greatly magnified. After Schaudinn.

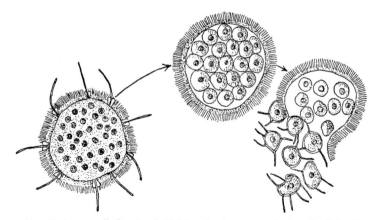

FIG. 3. A one-celled animal, *Trichospherium*, reproducing itself by the formation of spores. In this kind of reproduction the original "parent" ceases to exist as an individual, being completely dispersed into its offspring. Greatly magnified. After Schaudinn.

FIG. 4. The marine worm *Autolytus* reproducing itself by transverse fission. At *a*, a new head is forming, and the rear part of the worm will soon drop off to become a separate individual. Magnified. After Alexander Agassiz.

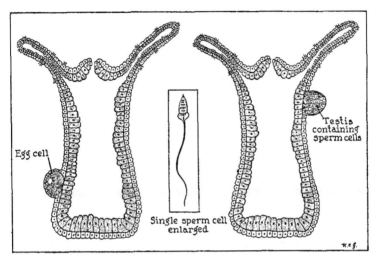

FIG. 5. Diagram of simple many-celled animal (*Hydra*) cut lengthwise to show the egg cell and the testis with its sperm cells. Compare with the photographs of the same subject, Plate II. From *Attaining Womanhood*, by George W. Corner, by courtesy of Harper and Brothers.

mals, notably the sponges and jelly fishes. In the common fresh-water polyp, Hydra, for example (Fig. 5 and Plate I), the buds develop from the side of the tubular parent and ultimately break away. An interesting development of this pattern is well seen in Obelia, a hydra-like animal often studied in biology classes, in which the bud is not exactly like the parent, but becomes a free-swimming medusa (jelly fish) which in turn produces a generation of polyps like the original hydroid.

In some of the sponges the buds or gemmules are formed internally and must await the death and decay of the parent before they can get free to begin their own career.

I have not space here to review all the modifications of this general sort that occur in the more primitive part of the animal kingdom. Some of them are decidedly bizarre. The process of budding can, however, be considered (with certain technical reservations) as merely a variation of the fundamental process of multiplication by fission.

H. G. Wells, Julian Huxley, and G. P. Wells, in "The Science of Life," summarize the whole subject of reproduction of living things when they say that "cleared of the complication of sex, reproduction is seen to be simply the detachment of living bits of one generation, which grow up into the next." Detachment of living bits of an animal is not always as easy, however, as in these primitive animals we have been considering. Obviously such processes as fission and budding can be effective only with relatively simple creatures. The more complex the parental animals, the more awkward for them to split in two or to produce buds. When, for example, there is a permanent hard shell outside the body, or a complicated skeleton inside, the animal cannot well divide itself in two. Animals with numerous special organs and tissues cannot readily form buds in which all the special features are represented. The new generation cannot take over the complex structure of its parent but must build its own body anew. When the parent detaches a bit of itself for the purpose of reproduction, that living bit must be a *germinal* organism, elementary and uncomplicated but able to grow rapidly and evolve itself into an adult like its parent.

I call attention to the fact that in this last sentence we have written the specifications of an Egg.

This idea was adumbrated long ago by an ancient balladist:

> *How should any cherry*
> *Be without a stone?*
> *And how should any wood-dove*
> *Be without a bone?*
>
> *When the cherry was a flower*
> *Then it had no stone;*
> *When the wood-dove was an egg*
> *Then it had no bone.*

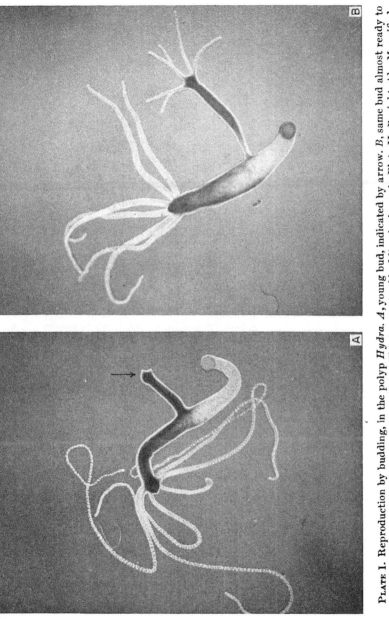

Plate I. Reproduction by budding, in the polyp *Hydra*. *A*, young bud, indicated by arrow. *B*, same bud almost ready to break off and become an independent animal. A still later stage of budding is shown in Plate II, *B*, right side. Magnified; size of animals about one-half inch long when fully extended. Photographs by P. S. Tice.

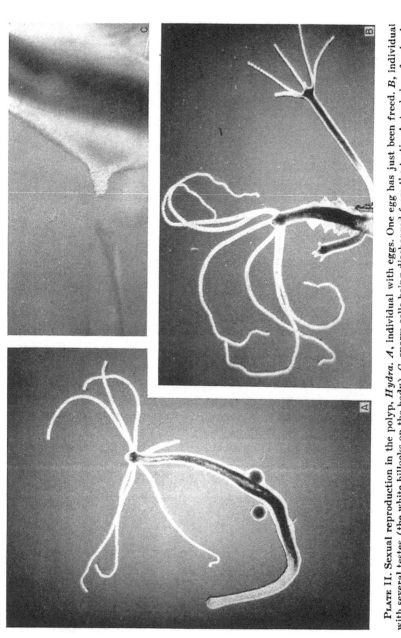

PLATE II. Sexual reproduction in the polyp, *Hydra*. *A*, individual with several testes (the white hillocks on the body). *B*, individual with eggs. One egg has just been freed. *C*, sperm cells being discharged from the testis. Actual size of animals same as in Plate I. Photographs by P. S. Tice.

The egg or ovum. We have already mentioned the simple fresh-water polyp Hydra as an example of animals that reproduce by budding. In this same animal, however, there is another kind of reproduction that occurs from time to time, in which the living part that is detached from the parent to form the new generation is not a bud, made of many cells and resembling the parent, but a single cell. As shown in the diagram (Fig. 5, left) and the photograph (Plate II, *A*) from time to time one of the cells in or near the surface of the animal enlarges very much and stores up materials with which it can be nourished for a while after it is cast off from the parent. This is the egg cell or *ovum*. The few cells that surround it where it grows on the side of the animal could be called an ovary (as we call the organ of similar function in higher animals) if it were worth while to dignify so simple and transitory a structure as the egg hillock of Hydra by considering it an organ. An egg, then, is a simple cell that is set aside by the parent and destined to divide into many cells and thus become an adult animal after the fashion of its kind. Seen in this light, reproduction by means of an egg is merely another case of reproduction by fission, in which the two living products of division are very unequal, the egg on one hand, the maternal animal on the other. If we compare a Hydra and its egg with an animal of a single cell, say a Cothurnia, that is going to divide, we see that the animalcule though an adult has also the function of an egg, for it can give rise, by division, to another animal body. In short, in one-celled animals the same cell must necessarily carry on all the functions of life, including reproduction; in many-celled animals the function of reproduction can be delegated to special cells.

The Scholastics debated which came first, the hen or the egg. Modern biology has an answer: they were contemporaneous; among protozoans the hen *is* the egg. Neither came first; they merely became distinguishable whenever it was (the record of evolution has some torn-out pages at this point) that an animal first became sufficiently complex to set aside a germ cell, specialized for reproduction. Had a scholastic philosopher been present on that prehistoric occasion the only question would have been which he noticed first—probably not the egg, because it was smaller than the rest of the animal.

The sperm cell. Hydras do not, however, always form eggs; half the time they develop not an ovary, but a testis, in which a few cells of the animal give rise by repeated division to a large number of very small sperm cells (diagram, Fig. 5, right; and photograph, Plate II, *C*). These cells can swim independently, when they are discharged into the water, by means of a motile tail with which each is provided. The sole function of such a cell is to swim until it meets an egg cell released from another Hydra and to enter it. When the egg is thus "fertilized" by union with a sperm cell, and then only, it begins to divide and ultimately to become a new Hydra. Herein we have the elements of sex, for this new polyp has two parents, which were not exactly alike in spite of their general similarity, because one of them furnished the egg and was temporarily at least a female; the other, which furnished the sperm cell, was temporarily a male.

THE MEANING OF SEX

No characteristic of man and the other animals is so fundamental, so completely taken for granted, as the existence of two sexes. It is the first fact the Bible mentions about the human race: ". . . male and female created He them." In every nature myth the animals enter two by two. In primitive song and story every Jack that cracks his crown has a Jill that tumbles after. Man that is born of woman finds it impossible to think of a race with only one sex, or to imagine other sexes than two. Nor does the biologist contradict this axiom; everywhere in nature he also sees two sexes.

Even in the lowest and simplest living things, in which it is fanciful to speak of male and female, there is (as we shall see) sexual mating or at least a process of renewal of life by the mingling of living substances.

In our day, however, science makes bold more than ever to question fundamental assumptions. The concept that space has three dimensions is as obvious as that animals have two sexes, but physicists do not hesitate to calculate in four, five, or n dimensions. We may boldly ask, therefore, why sex is necessary at all; or why there are not several sexes. If living things must mate in order to reproduce why could not nature have arranged some other system, for example a state of sexual relativity, in which an individual might be (without any change in itself) male with respect to one potential mate, female with respect to another? or it might take part in reproduction in response to another of its species, neither of them being either male or female. Such conjectures are not more fantastic than the concepts of mathematical relativity, with their notions of a warp in space, and of an expanding universe. Similar questions have indeed long been asked by the poets and philosophers. John Milton vigorously states an unfavorable view of the two-sex system:

> *O why did God,*
> *Creator wise, that peopl'd highest Heav'n*
> *With Spirits Masculine, create at last*
> *This noveltie on Earth, this fair defect*
> *Of Nature, and not fill the World at once*
> *With Men as Angels without Feminine,*
> *Or find some other way to generate*
> *Mankind? This mischief had not then befall'n.*

PARADISE LOST.

and Sir Thomas Browne, the famous physician philosopher, tough-minded as he was on many subjects, was personally squeamish about the whole matter of reproduction by physical contact of the sexes:

> I could be content that we might procreate like trees, without conjunction, or that there were any way to perpetuate the World without this trivial and vulgar way of union.
>
> RELIGIO MEDICI.

In illustration of these roving thoughts of poet and scientist, let us return for a moment to the sea-born Cothurnia. When such an animalcule reproduces itself by division, we find it handy to call the new animal a "daughter cell," but this is only a figure of speech. The offspring is even closer than a child; it is more truly a twin of the cell that produced it, for their relationship is exactly like that between a pair of human identical twins, which arise by the splitting of one cell, i.e. the human egg. Presently the parent Cothurnia will give off another "daughter cell"; will that be niece or sister of the first? And if a Cothurnia is sister to its mother (i.e. the cell that produced it) is it not equally sister to its grandmother . . . and so on, as far back as the line was reproducing by simple division? We have here indeed a situation in which the terminology of the human bisexual family tree breaks down, for all the offspring of a single cell that reproduces without mating are related together even more closely than the members of a human family. They all have identically the same heredity, and are all "twins" one of another, though they may number thousands and represent many "generations." For this kind of group the biologists have had to invent a new name; they call it a *clone*.

Such a clone, being a group of cells that all came from one cell, may be considered something like the body of an individual many-celled animal. That too is a group of cells that all came from one cell. Like an animal body the clone seems to have its phases of youth, maturity, and old age. After many divisions it becomes old; the vigor of its members diminishes. Individual animals become abnormal and enfeebled, and the rate of fission slows down. The clone is in danger of extinction. It needs a shake-up, which it cannot get through the process of complete inbreeding (or rather, lack of outbreeding) by which the clone develops.

Conjugation or mating. This seems to be the reason that even in such simple animals a process much like sexual union occurs. Fig. 6 shows the mating of the one-celled animal *Scytomonas.* Animals of this species are not attached, like Cothurnia, but swim about in the water. As we watch them under the microscope, two individuals that are going to mate swim near each other, come into contact and actually cohere side to side. One of them loses its whiplike flagellum. The protoplasmic substance of which they are made becomes continuous from one animal to the other, and the nuclei move toward each other and unite. In this particular species the conjugated animal then becomes dormant for a time, but ultimately resumes

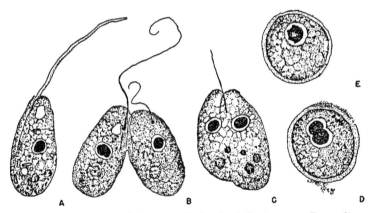

Fig. 6. Conjugation of the one-celled animal *Scytomonas.* Proceeding from *A*, which shows a single individual, through *B*, *C*, *D*, and *E*, two cells are seen to join, fuse their nuclei (the dark round objects) and unite into one cell. This single individual remains dormant for a time but ultimately becomes active again. Greatly magnified. After Dobell, simplified.

activity and becomes the parent of a new clone. This process is very much like fertilization of an egg by a sperm cell in higher animals, and we can get from it a better understanding of the meaning of sexual union. For example, the late Professor H. S. Jennings (a brilliant predecessor of mine in the Vanuxem Lectures) with his fellow workers has investigated the mating habits of a very well known one-celled animal, Paramecium. They have discovered the remarkable fact that two Paramecia of any one clone will not conjugate with each other. The animal must find a mate not closely related to it. By studying an immense number of animals of the species *Paramecium bursaria* the investigators found that the whole population of the species is distributed into several "mating types" such that an individual of one type will mate with one of another type, but not with one of its own.

Conjugation is therefore a kind of outbreeding. To use a figure of speech taken from higher animals, it introduces "new blood" into the family, which causes an internal re-arrangement of the cell materials and gives the race a new start. The "mating types," it will be noted, are not in any strict sense different sexes. As Jennings points out, in one of his examples, types A and B will mate together and type C will mate with either of them. Such an observation shows that type C is not of a fixed "sex." The situation indeed is one of relative sexuality, such as we cited above when we were trying to imagine other ways of reproduction Nature might have tried rather than the bisexual method that universally characterizes higher animals.

In some other one-celled animals, however, there seem to be only two mating types, and in many species the two conjugating types are actually somewhat different in appearance. This is getting closer and closer to bisexuality in the strict sense.

It is probable that conjugation or something like it goes on in every kind of animal, although the details are not always the same, and there are puzzling and obscure cases awaiting solution. At any rate the situation is clear in the higher animals, which always reproduce by the union of an egg cell with a sperm cell. Just why the whole animal kingdom, except a few of the lowest and simplest creatures, settled down so completely to the egg-and-sperm system, we can only guess. Perhaps in some early ancestral animal, not originally bisexual, it happened that some of the reproductive cells were unusually well stored with nutritive substances. This would be an advantage, for it would help to tide the embryos over the earliest stages of their development before they could feed themselves. Such cells would, however, be sluggish, and the chance of two of that kind meeting would therefore be reduced. A germ cell that happened to be lighter and more mobile would be more likely to meet the relatively sluggish cell. Once started, such a trend toward two types would progress and become fixed, by the familiar Darwinian process of survival of the fittest, and ultimately we should arrive at the characteristic arrangement, namely large eggs laden with food (yolk) and small active sperm cells.

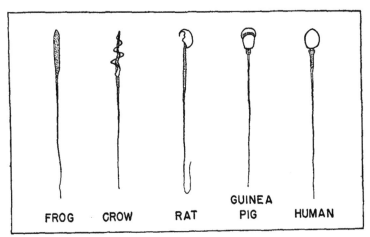

FIG. 7. Sperm cells of various animals and man. Greatly magnified.

Fertilization of the Egg

The fertilization of an egg by a sperm cell is one of the greatest wonders of nature, an event in which magnificently small fragments of animal life are driven by cosmic forces toward their appointed end, the growth of a living being. As a spectacle it can be compared only with an eclipse of the sun, or the eruption of a volcano. If this were a rare event, or if it occurred only in some distant land, our museums and universities would doubtless organize expeditions to witness it, and the newspapers would record its outcome with enthusiasm. It is, in fact, the most common and the nearest to us of Nature's cataclysms, and yet it is very seldom observed, because it occurs in a realm most people never see, the region of microscopic things. It is, moreover, in most animals we are likely to see, a recondite event, occurring in ponds or the sea, in the forest or the earth, wherever the creatures lay their eggs. In mammals and birds the fertilization is hidden in the depths of the body. Nor indeed are all eggs suitable for study; they may, for example, be opaquely loaded with pigment like those of the frog. Such eggs may, of course, be killed and cut into thin slices for microscopic study, and the process of fertilization has thus been observed step by step in the prepared eggs of many species, but only a few biologists ever see the whole continuous process of union of a living egg with a sperm cell.

It need not be so rare a sight, however, for anyone who will go to a seaside laboratory in summer can witness it. The sea urchins, starfish and sand dollars which inhabit our coasts almost seem especially created to reveal the process of fertilization with utmost clearness. While writing this chapter I have before me a sketchbook made while a college student, working at the U.S. Fisheries laboratory at Beaufort, North Carolina, where I studied with amazement the finest of all these marine eggs, those of the white sea urchin, *Toxopneus-*

tes variegatus, first described by Louis Agassiz, and introduced to experimental biology by the cytologist Edmund B. Wilson. A related species is shown in the beautiful and instructive photographs here presented (Plates III and IV), the work of Dr. Ethel Browne Harvey of Princeton, to whom I am deeply grateful for the opportunity to use them. They were made at Woods Hole from a common northern sea urchin, *Arbacia punctulata*.

The male and female sea urchin deposit their sperm cells and eggs, respectively, directly into the sea. For purpose of study, however, it is quite readily possible to remove the germ cells from the animals before they are spawned and to bring them together in a dish under the microscope. The observer cuts open the spiny shell of a female urchin and pulls out the ovaries, slits them and catches in a dish of water the hundreds of beautiful glass-clear spheres, 0.9 mm. (0.037 inch) in diameter. A male sea urchin yields its testes, from which exudes a fluid milky with microscopic particles, each of which is a wriggling, dancing sperm cell—a tadpole-shaped object with a lance-shaped head about 0.07 mm. (0.003 inch) in length and a long tail. When the sperm cells are mixed with the eggs, they swim about rapidly until they touch the eggs, to which they adhere, several sperm cells about each egg, trying to push into its substance. When one sperm cell has actually penetrated the egg (Plate III, *A*, *B*) it causes the surface of the egg cell to be rapidly congealed into a thin membrane, something like the scum on a cup of cocoa. By this means, the other competing sperm cells are effectively prevented from entering.

Meanwhile the observer will have noticed the *nucleus* of the egg (Plate III, *A*), a rounded body about one-sixth the diameter of the whole egg cell, eccentrically placed near one edge and enclosed by a delicate nuclear membrane. This nucleus is the goal toward which the sperm cell is moving, and the object of the whole process is to secure fusion of the sperm cell with the egg nucleus. The tail of the sperm cell is broken off and left behind. The head now advances through the egg, swelling slightly as it goes. In the egg substance a star-shaped aster or region of stiffened egg substance appears and travels with the sperm nucleus. The egg nucleus advances to meet the sperm and within ten minutes from the time the sperm cell enters the egg, the two nuclei have united and blended their substance (Plate III, C). The egg is now fertilized; it immediately prepares to divide, and while the observer watches, entranced with the smooth inevitability of these events, the divisions follow one another every twenty-five minutes, so that one cell becomes two, the two become four, eight, sixteen, thirty-two, and so on until the embryo is a mass of small cells looking like a mulberry. All these events are shown in Plates III and IV. We need not follow the fertilized egg through the subsequent complicated changes and metamorphoses by which it becomes an adult sea urchin.

The meaning of fertilization. If the eggs are left alone in the dish, they do not go ahead by themselves and turn into sea urchins. Development cannot begin until after the entrance of the sperm cell and the fusion of the nuclei. Evidently the sperm cell in some way is necessary to set off or stimulate division of the egg. A great step toward understanding what happens was made by Jacques Loeb in 1899 and 1900. Loeb had been studying the effect upon life processes of changing the amounts of certain minerals which are present in living tissue. By increasing or decreasing the concentration of magnesium or calcium in sea water, for example, he could speed up or slow down the rate of division of fertilized sea urchin or starfish eggs. This led him to try dilute solutions of magnesium chloride on the unfertilized eggs, completely free from sperm cells. Eggs so treated began to divide and when carefully handled often went on to form complete larvae. In later

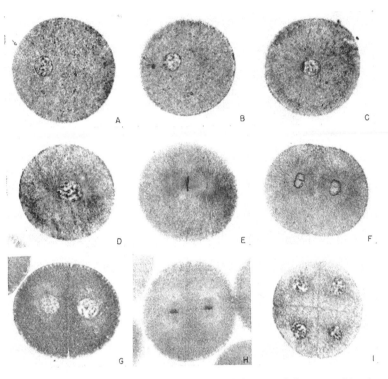

PLATE III. Fertilization and segmentation of the egg of the sea urchin *Arbacia*, as seen in eggs sectioned and stained for microscopic examination. *A*, sperm cell about to enter egg. *B*, sperm nucleus (small black-stained object) approaching egg nucleus. *C*, sperm nucleus (black object) fuses with egg nucleus, 10 minutes after entry of sperm cell into egg. *D*, nucleus of fertilized egg begins to divide. *E*, division in progress. The black-stained objects in a row at center are the chromosomes. *F*, nuclei of the two daughter-cells reforming. *G*, stage of two cells. *H*, the first two cells now divide again. *I*, stage of four cells. All magnified 375 diameters. Prepared and photographed by Ethel Browne Harvey.

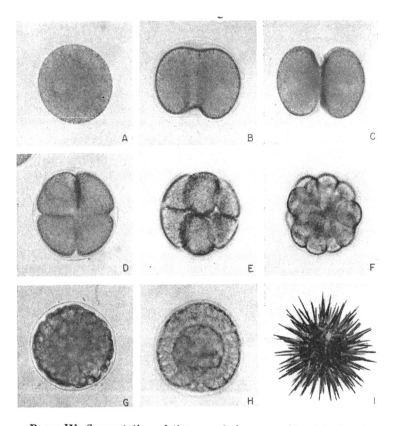

PLATE IV. Segmentation of the egg of the sea urchin *Arbacia*, photographed while living by Ethel Browne Harvey. *A*, fertilized egg. *B*, beginning of segmentation. *C*, two-cell stage. *D*, four cells. *E*, eight cells. *F*, sixteen cells. *G*, "mulberry" (morula) stage. *H*, hollow embryo (blastula). *I*, young adult sea urchin. *A* to *H*, magnified 290 diameters; *I*, almost natural size.

experiments by others, such fatherless young have actually been raised to adult life, and to all appearances were normal specimens of their kind.

Lest this experiment should seem to disparage the importance of the father, we should mention that the contrary experiment also succeeds. If an egg is cut into two pieces, one of which has no nucleus, and the latter is then entered by a sperm cell, it too will divide and become an embryo, though admittedly not as often as in the other, less drastic experiment. In this case the embryo is motherless, from the standpoint of heredity, for it has no egg nucleus in it. This shows that egg stuff, to develop, must have a nucleus and requires to be stimulated, but either an egg nucleus or a sperm nucleus will do. We shall see later, however, that for reasons that concern heredity it is decidedly better for the offspring to get its nuclear material from both parents, as normally happens.

This remarkable experiment of *artificial parthenogenesis* ("virgin generation") as it is called, has been repeated on many kinds of animals, and it has been found that not only magnesium solutions, but quite a number of different stimuli will start division of the eggs. Exposure to sperm cells of other species, extracts made from dead sperm cells, various dilute acids and alkalies, sudden cooling, heating, shaking, or pricking the eggs all can be used to initiate development in one species or another. Mere staleness will cause the eggs of some animals to divide. In all probability these diverse stimuli produce some sort of common effect on the cell substances, setting up internal changes (not as yet well understood) that start the processes of division and growth of the egg. The point of interest for us is that when the sperm cell acts upon the egg in this way, it is exerting a merely physical or chemical effect. The fact that it is itself a living cell is more or less incidental. The egg contains all the essential elements for the production of a perfect animal, and needs only to be given a start.[1]

The eggs of mammals, including the human species, probably do not differ in this respect from those of sea urchins and starfish. Because they are far harder to get at and less resistant to handling and exposure, experimental study has not progressed very far. In very recent years, Gregory Pincus and his associates at Clark University have worked with rabbits, using an experimental method in which the unfertilized egg was subjected to drastic cooling while passing through the oviduct (Fallopian tube). A few such eggs, subsequently replanted into other rabbits, are said to have formed embryos, to have been born, and to have grown to normal adult life. H. Shapiro of Philadelphia has still more recently reported starting development of the rabbit's egg by drastic refrigeration of the whole body of the female rabbit, but up to the present none of these eggs has developed beyond the earliest embryonic stages.

I hasten to add that under ordinary circumstances, when there is no meddling by an experimenter, mammalian eggs live in a perfectly conditioned environment. The temperature and all other conditions to which they are subjected in the

---

[1] In a brief chapter like this, in which I am deliberately selecting those features of the natural history of reproduction which best lead up to the higher animals, it is not possible to follow out all the ramifications of the subject. Life processes are so richly varied that every general statement calls for a bill of exceptions. There are, for example, many animals that can produce parthenogenetic eggs, i.e. eggs that develop spontaneously without fertilization. This is the case in a great many insects. In some of these instances, no doubt, a stimulus akin to that of fertilization is furnished by natural conditions, such as high temperatures, desiccation, or chemical changes within the egg, but in others there is no known special stimulus. Indeed, when we reflect that a tendency to propagate by division is innate in almost all animal cells, the wonder is that in most species the eggs do have to be stimulated in order to develop.

Incidentally, even in the insects and other animals with parthenogenetic generations, sexual reproduction always occurs from time to time to rejuvenate the line and start new clones, just as in one-celled animals. In all vertebrate animals sexual reproduction is obligatory.

body are so closely regulated that parthenogenesis of the sort described by Pincus would not be possible.

Heredity. We have not yet told the whole story of fertilization. Mere stimulation of an egg to develop, necessary as it is, is not all the sperm cell does. It has another vastly important task, which is to carry into the egg the male parent's contribution to the heredity of the offspring. Packed in the nucleus of the egg and in the nuclear head of the sperm cell are the submicroscopic chemical particles that control the inherited characteristics of the species; when the sperm cell unites with the egg, the nucleus of the fertilized egg acquires an equal share of this controlling material from each parent. When the egg divides, these determiners are distributed to the daughter cells at each division and thus are carried into all the cells of the embryo. This is the way the sperm brings in "new blood" and rejuvenates the cell lineage, as happens in the one-celled animals by means of conjugation. In this way, moreover, two family lines are blended, and special traits of bodily build and behavior are exchanged and distributed, so that the young are never quite identical with either parent.

This blending and assortment of hereditary characters is the object and goal of sexual reproduction. Whatever else follows in this book is merely the story of the arrangements and devices of nature to assure the meeting of egg and sperm cell and to protect the embryo that they produce.

How the determiners of heredity that can shape the whole body of a man or woman, and then bequeath themselves to another generation, are packed into the small compass of egg and sperm cell, how they are distributed to the cells of the body by processes of almost geometrical precision, how they can be traced and how they guide the building of the body—this is the subject matter of the science of genetics, one of the grand divisions of modern biology. It is not to be a theme of the present book. Those of my readers who have studied biology have mastered at least the rudiments of genetics and the lore of the chromosomes. Unfortunately, this important and beautiful science is almost impossible of explanation to those who have not seen animal and plant cells through the microscope. Several writers have made brave attempts to do so, and the reader is referred to their books.[2] For us let it suffice that egg and sperm cell join; we shall not attempt here to see what goes on within them.

Meeting of the germ cells. The eggs and sperm cells of such simple and small animals as Hydra are discharged directly into the water, and the sperm cell swims to the egg. As animals become larger and more complicated, the *gonads* (ovaries and testes) are built more deeply into the body. Some sort of opening or channel to the surface is then provided. The ovaries and testes of sea urchins, for example, open through the shell by small pores.

When fertilization depends upon the chance meeting of eggs and sperm cells, or upon such uncertain aids as tides and currents, there is obviously a great risk of failure to make contact. To compensate for this, and also for the subsequent high loss of embryos, due to enemies and unfavorable conditions, an enormous excess of germ cells is usually produced. There would obviously be greater economy and safety if some arrangement were made to bring the germ cells together or to put them near each other in the first place. A fantastic variety of such arrangements is seen in nature. Animals that are sluggish like the sea urchins and starfish, or actually fixed in position, like most shellfish as well as the sponges, corals and many ascidians (of which the "sea squirts" are examples) are often aided, as we said above, by tides or other currents. In higher plants, which are not only rooted to the

[2] H. G. Wells, Julian S. Huxley and G. P. Wells, *The Science of Life*, London, 1929; Charles R. Stockard, *The Physical Basis of Personality*, New York, 1931; Alan F. Guttmacher, *Life in the Making*, New York, 1933; A. M. Scheinfeld and M. D. Schweitzer, *You and Heredity*, New York, 1939.

ground, but have male germ cells that cannot move of their own accord, the winds or the insects transport the pollen.

Most of those animals that are free to move mate by propinquity. They can at the very least deposit their eggs and sperm cells at the same place. This is the case in many fishes, in which the male and female place themselves close together when they spawn, so that the sperm cells are deposited upon the eggs. Frogs and toads provide an even better chance of contact between the germ cells, for in the mating season the male instinctively clasps the female with his fore limbs and the two animals remain in close contact for days, until the eggs are discharged, whereupon the sperm cells are deposited directly upon the eggs. In the tailed amphibians (such as newts and salamanders) sperm cells are not discharged externally at all. Like all other vertebrates below the mammals (i.e. fish, amphibians, reptiles, and birds) these tailed amphibians have a combined cloacal passage into which both the intestinal and the genital canals open. The openings of the cloacas of the male and female are placed so close together in mating that the sperm cells pass directly from one to the other. They then pass up into the oviduct and fertilize the eggs there. Much the same process occurs in many birds, for example the common fowl and their kin.

An obvious advance is the development of an organ for direct transmission of the sperm cells. The elasmobranch fishes (sharks and rays) possess specially modified anal fins, called claspers, which are grooved so that the seminal fluid containing the sperm cells is guided along them from the cloaca of the male to that of the female. This particular method of solving the problem is, of course, not available to land animals, since they have no fins. In snakes and lizards there are saclike branches of the cloaca that can be turned inside out and protruded into the cloaca of the female, carrying with them the sperm cells. In other reptiles, namely turtles and crocodiles, and in many birds the final solution was attained and has become standard in mammals. This is a special male organ, the penis, adapted to insertion into the female genital tract. In female mammals the lower end of the genital canal is expanded into a special canal, the vagina, which receives the penis. The sperm cells can thus be safely placed well within the reproductive system of the female. In both sexes, in mammals, the intestinal outlet is separate, leaving the genital outlets (penis and vagina respectively) associated only with those of the urinary system.

In the higher vertebrates, then, the eggs leave the ovary and pass down the egg ducts. If mating occurs, sperm cells are put into the vagina (or into the cloaca in reptiles and birds) and travel upward to meet the eggs in the oviduct and fertilize them there. What is to be done next with the eggs? Turtles coat them with a parchment-like shell (secreted by the lower end of the oviduct), lay them, and bury them in the sand. Birds provide a hard shell and put them in a nest. Mammals do much more for their fertilized eggs—they keep them in the mother's body and develop them there. We shall have to study in later chapters the elaborate arrangements necessary for this process of *gestation*.

Gestation

Although the development of the young within the mother's body is characteristic of the mammals, and is most highly perfected in that order of animals, it is by no means unknown in lower orders. In some of the mollusks, for example, the embryos are kept within the shell of the mother until their development is well advanced. The European oyster thus long retains its embryos in the gill chamber. The viviparous fish which have become popular in home aquariums in recent years raise their young in the oviduct. Some fish actually retain them in the ovary itself. The young fish, which are very small, live on the yolk that was in the egg from which each one sprang, and in some species probably absorb nourishment from the tissues of the mother, but they are not actually attached to her. In some species of dogfish, the lower part of the egg duct is expanded into a special chamber. In this the embryos are retained. The lining of the chamber is thrown into a mass of finger-like projections, between which grow similar projections from the belly wall of each embryo. Nutriment brought to this zone of interlacement by the blood vessels of the mother filters through the coverings of the two sets of projections into the blood vessels of the embryo, as moisture filters into the roots of a tree. Much the same arrangement prevails in the viviparous snakes.

In mammals the brood chamber is more than a mere dilatation of the oviduct. It becomes a special organ, the *uterus*, which has thick muscular walls, to enable it to withstand the distention produced by large embryos during weeks or months of development, and afterward to expel them when the time comes for their birth. The attachment between mother and child (the *placenta*, to be described more fully later) becomes very intimate and very effective in transmitting nutritive substances to the embryo and carrying waste products away. Instead of being thrust into the outside world as an unprotected egg, the mammalian infant is sheltered and nourished in the uterus for a long time—three weeks in the mouse, four months in the pig, nine months in man, two years in the elephant. Even after so long a period of gestation, when it enters the world it is still dependent upon the body and secretions of the mother, for it cannot do without the milk she provides for its nourishment.

A modest word about the father may be in order at this point. It will perhaps seem from our sketch of his biological function that in all the various races of animals his duty and usefulness are done when he has put his sperm cells where they can reach an ovum, serving thus to set the mechanism of the egg into action and to contribute his equal share to the heredity of his offspring. The rest takes care of itself in lower animals, and in the higher orders seems a task for the female alone. There is indeed one species in which the male animal plays no other part at all in life than this—the marine worm *Bonellia*, famous in biological lore, in which the male is nothing but a very small parasite on the gills of the female. She carries this petty creature about with her for the sole purpose of getting the eggs fertilized. Yet he cannot wholly be dispensed with, however brief his moment. In human history a not dissimilar career has been that of certain prince consorts of masterful queens.

Such is not altogether the case in mammals. The very fact that gestation is a heavy burden, putting the female at a disadvantage in the struggle of life, while she is carrying her young and afterward mothering them, gives the male parent another task—that of protector and leader of the family. This finds biological expression in the fact that in almost all mammals the male is larger, stronger, and fiercer than the female (Rudyard Kipling to the contrary notwithstanding). In the human race the mother's burden is heaviest of all, and by that very fact the father becomes again biologically useful to his offspring during the long period of gestation and infancy, as guardian and provider of food and shelter.

Meaning of Sex for Human Beings

The gist of our preface to human reproduction is that our own species and most of the others, high and low, reproduce themselves by the production and union of eggs and sperm cells. To get this essentially simple task done in the highest animals requires the functioning of an elaborate set of organs. In order to tell the whole story of reproduction in man and the higher animals we shall have to discuss:

The anatomy of the ovary and testis and the formation of eggs and sperm cells,

Transportation of the egg from ovary to uterus, and of the sperm cell from testis to the egg,

The union of the germ cells,

Development of the fertilized egg,

The attachment of the embryo to the mother, and its nutrition,

Birth and the nursing of the young.

These are complicated matters, which must be timed so that each stage fits into the next. What goes on in one organ must be coordinated with events in another. The chemical environment of the egg and the sperm cells must be kept in adjustment to their needs; muscle cells in the oviduct, uterus and the male reproductive system must be ready to act when required; the lining of the uterus must be prepared for the embryo; in short, a whole complex system of organs and tissues must work as a unit.

The body has two important ways of linking the action of its separate organs. One of these is the nervous system, through which run innumerable signals connecting the organs of sensation and motion and regulating many functions of the internal organs. The other, with which we are much more concerned in this book, is the system of the chemical messengers or *hormones*. A hormone is a chemical substance made in one of the special glands called ductless glands or glands of internal secretion, among which are the pituitary, thyroid, parathyroid, adrenal, and parts of the pancreas, which do not discharge their product through a duct to the outside of the body or into another organ, as for example do the sweat glands, the salivary glands, the liver and the kidney. Instead, these endocrine glands (for such they are also called) put their respective secretions into the blood as it courses through the blood vessels which pervade their substance. The hormones are thus carried all over the body and reach the various organs and tissues which each of them is respectively destined to affect. The hormone of the thyroid, for instance, influences the utilization of oxygen by the tissues; the pancreatic hormone (insulin) regulates the combustion of sugar; adrenin affects the blood pressure by causing the arteries to contract. Just how each ductless gland produces its own special secretion, and why certain particular tissues, and these only, respond to a given hormone, are questions which must be solved by the physiologists and the chemists of the future.

The ovaries and the testes are also glands of internal secretion, or to put the case more precisely, they include such glandular tissue among their complicated make-up. The hormones made in the sex glands perform the function of linking the action of the various organs of the genital system, timing and regulating their activities. What these hormones are, where they are produced, and how they act to accomplish the bodily tasks of human reproduction, is the theme of this book.

One more item should be added to the list of reproductive activities cited above. This is the sexual urge, the totality of impulses that serve to bring the sexes together for mating. It is the most important coordination of all, for without the union of the sexes all the other intricate processes are useless. It, too, is partly regulated by the hormones, but we know too little about this as yet to discuss it profitably here. The way of an eagle in the air, the way of a man with a maid . . . are still in part beyond the reach of science. In view of the fact that we are still ignorant of the means by which the simplest one-celled animal is impelled to conjugate with another of its kind, we can only wonder at the complexity of sex psychology in the higher animals, and at all the lures that nature has provided to insure the union of the sexes. What marvels of color and fragrance, bird song and firefly radiance, have been lavished to this end! and for mankind what emotions are bound up with it, of young romance and mature devotion, hope and fear, selfishness, slyness and cruelty. To the fanatic, sex is a snare of the devil, to a Casanova a heartless game; to Stephen Dedalus as a young man it was torment; to some happier lad, it is a rosy dream. All these have much to gain by seeing it also, with the biologist, as part of the inevitable process of animal life. To understand is not to demean ourselves, nor to rob the human heart of virtue and the love of beauty.

THE HUMAN EGG AND THE ORGANS THAT MAKE
AND CARE FOR IT

"*All those parts of the* Hen *which are designed to* Generation, *namely the* Ovary, Infundibulum, *the* process *of the* Womb, *and the* Womb *itself, and the* Privities: *and also the* scituation, fabrick, quantity, *and* Temper *of all these, and whatsoever else relates thereto: they are all inservient, and handmaids either to the procreation of the* Egge, *or to its Augmentation, or else to* Coition, *and fertility received from the* Male, *or to the* foetus: *to which they conduce either necessarily or principally, or as a* Causa sine qua non, *or some other way to the better being. For there is nothing made either vain or rash in all the operations of* Nature."—WILLIAM HARVEY, *Anatomical Exercitations Concerning the Generation of Living Creatures*, 1653.

·§ CHAPTER II §·

*THE HUMAN EGG AND THE ORGANS THAT MAKE
AND CARE FOR IT*

WHAT can we say of the ovary, an organ so remarkable that it is able to produce the Egg? The human ovaries (Plate V, *ovary*) are insignificant-looking whitish, tough bits of animal tissue each about the size of a small walnut[1] hanging from the broad ligament at the back of the pelvic cavity, beside the uterus. They were frankly a puzzle to the ancient anatomists, who never imagined that mammals have eggs and could not have seen the tiny human eggs in the ovary anyway. The hen's egg and ovary they understood much better, for they could see in the ovary the developing yolks of various sizes and thus they perceived the connection between the organ and the ova it produced.

In another place[2] I have told the story of the discovery of the mammalian ovum—how in 1672 the brilliant young Dutchman Regner de Graaf described what we now call *Graafian follicles* or simply *ovarian follicles*, round egg-chambers, filled with fluid, that he saw in the ovaries of cows, sheep, swine, rabbits, and women (Plate VI, *B, C, E*). Familiar, of course, with the eggs of birds, he thought each follicle in the mammal was an egg. He was surprised, however, that he could not find such "eggs" after they left the ovary, for in large animals the mature follicles are as big as peas or small cherries, and ought to bulge the oviducts as the hen's eggs do. After long and futile search for the real egg by many

[1] The human ovary is about 3 centimeters (1.25 inches) long. The ovary of a mouse is hardly bigger than a pinhead, while that of a whale weighs 2 or 3 pounds; but their eggs are all about the same size, as we shall see.

[2] George W. Corner, "The Discovery of the Mammalian Ovum," *Mayo Foundation Lectures in the History of Medicine*, 1930.

{ 33 }

workers, Karl Ernst von Baer solved the puzzle in 1827 by finding that the actual egg is a very small speck inside the follicle, too small to be picked out by the unaided eye. If we take one out of its place in the ovary and put it by itself in a dish of clear water in a bright light it can just barely be seen.

With modern instruments we can make a very thin slice of an ovary, put it under the microscope, and photograph it. Studying the slide from the surface down (Plate VII, *B*) we see a layer of covering cells (germinal epithelium), appearing as a dark line at the top of the photograph, with a vague zone of connective tissue beneath it, a zone of egg cells in reserve (*c*) and follicles (*f*) of various sizes, each containing its egg.

The follicle (see again Plate VII, *A*) is lined with a layer which looks granular under low magnification, because it is made up of small cells. Just outside this *granulosa* layer is a thin layer (called *theca interna*) of larger cells well supplied with blood vessels. This layer, so slight that it would hardly be noticed, except by an experienced microscopist, is of great importance because in all probability it is the source of the *estrogenic hormone* which (next to the egg) is the most important product of the ovary.

How the eggs are formed in the ovary is an unsettled problem. We know that large numbers of them are produced by ingrowth from the surface cells of the ovary before birth. In a newborn baby girl there are already thousands of egg cells, many more than she can possibly need when she grows up. Many anatomists think that these original eggs, present at birth, furnish the supply for life; in other words, that no new ones are formed after birth. This means that the infant in arms has already set aside her contribution to the heredity of her children, and if perchance she has her last baby at forty years of age, that particular egg will have waited all those forty years for its opportunity to develop. This is

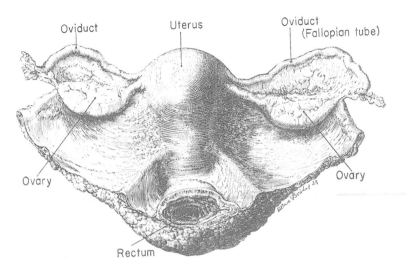

Oviduct   Uterus   Oviduct
(Fallopian tube)

Ovary   Ovary

Rectum

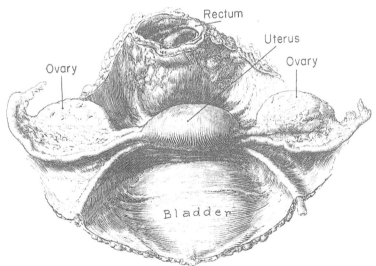

Rectum

Ovary   Uterus   Ovary

Bladder

PLATE V. The human female reproductive organs, drawn by a famous
medical illustrator, the late Max Broedel. *Above*, viewed from dorsal
(rear) aspect. *Below*, viewed from above, as seen by surgeon looking into
the pelvis. One-half natural size. From the Carnegie *Contributions to
Embryology*, by courtesy of George B. Wislocki.

Tab XV

somewhat staggering, but not altogether preposterous, for we have good reason to think that some other very important cells, e.g. the chief cells of the brain and spinal cord, last a whole lifetime without replacement. Other anatomists think they can see scraps of evidence that new crops of eggs are continually being formed in the ovary of the adult woman, and that these new eggs are those which are shed from the ovary in adult life. It is well known that the male germ cells are formed anew continuously in adult animals (see Chapter IX). The question, as regards the ovary is (for technical reasons) much more difficult to solve than it might seem. I have myself studied it quite seriously in a large collection of monkey ovaries, and thus far have not seen good evidence that new formation of eggs occurs. For this reason I adhere cautiously to the old view until new evidence is brought forward.

At any rate, some of the eggs in the reserve zone are from time to time selected to proceed to maturity. Such an egg sinks deeper into the ovary. The cells about it multiply to form a thick mass, which soon hollows out to form a small follicle. As the cavity enlarges, the egg is left in its little hillock at one side (Plate VII, *A*, *B*, *C*).

Discharge of the egg. The follicle continues to grow and to occupy more and more space, slowly shoving aside neighboring tissues within the ovary. Finally it enters a period of very rapid growth, so that its volume doubles in a few hours. Plate VIII shows the growth of the follicle in the rat. The human follicle becomes 12 or 15 millimeters (0.5 to 0.6 inch) in diameter when fully developed and occupies at least one-fourth of the whole volume of the ovary. As it enlarges, it pushes its way to the surface. The wall of the follicle next the

PLATE VI. De Graaf's original illustration of the Graafian follicles, in the ovary of the cow, from *De organis generationem inservientibus*, 1672. A large follicle is shown at *B*, smaller ones at *C, C, C, C*. *E* is a large follicle dissected away from the ovary. The lower portion of the figure shows the oviduct (Fallopian tube) with its funnel-like expansion. Approximately natural size.

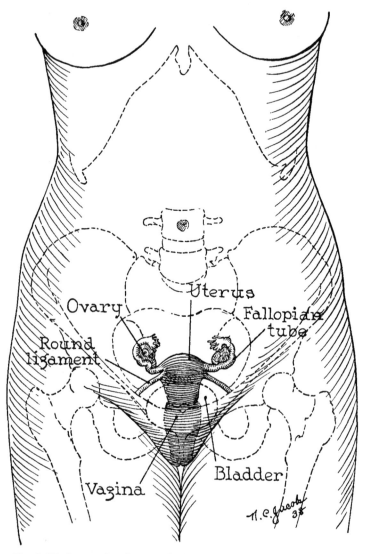

FIG. 8. The human female reproductive system. Dotted lines indicate the position of the pelvis and other bones. From *Attaining Womanhood*, by George W. Corner, by courtesy of Harper and Brothers.

surface and the overlying capsule of the ovary become thinner
and thinner. The actual rupture of the follicle through the
thinned-out region has been observed repeatedly in rabbits
and sheep, following the lead of Walton and Hammond of
Cambridge, England (1928), and has been photographed in
motion pictures by Hill, Allen and Kramer of Yale Medical
School. It happens that in the rabbit these events in the ovary
can be timed very closely. The investigators anesthetize an
animal which is about to ovulate and expose one of its ovaries
under warm salt solution. Several ripening follicles are seen.
Watching or photographing one of these, they see the thin
exposed wall of the follicle weaken still more, until it bulges
to make a little bleb. Meanwhile the cells about the egg on
the inner wall of the follicle have loosened up so that the egg
is nearly free from its attachment. Finally the bleb rips open
and the contents of the follicle is expelled, carrying the egg
with it. In motion pictures, the ejection of the contents looks
like the slow rise and fall of a geyser; it is not explosive, but
actually rather a gentle occurrence. In animals like the rabbit, which produce several young at one time, the individual
follicles of one batch rupture within a few minutes of each
other.

The egg. The eggs of mammals, seen under the microscope,
are beautiful little spherical objects, consisting of a round
mass of cellular material surrounded by a transparent zone
or membrane (Plate VII, *D*). The eggs of all the higher
mammals thus far measured have been not far from 0.1
millimeter (0.004 inch) in diameter. Mouse and rat eggs are
a little smaller (0.075 mm.), those of dog, cow, and human a
little larger (0.140 mm.). Dr. Carl G. Hartman suggests a
striking comparison by which we can appreciate their size,
relative to more familiar objects: "Scatter a pinch of sea
sand on a piece of black paper—the smallest grain visible to
the naked eye is of the order of magnitude of the cow's egg."

As Hartman calculates, it would require about 2,000,000 eggs to fill an average sewing thimble.

We know fairly well the appearance and size of the human ovum while it is still in the growing follicle of the ovary, but our information about the fully mature egg (that is, during the last hours before it leaves the ovary, and while it is in the oviduct) is derived from a mere handful of specimens, about ten or twelve in all, that various investigators have been able to obtain. The Rhesus monkey has yielded to science a somewhat larger treasure of eggs. If anyone wants monkey eggs in market lots, they might be furnished for two or three thousand dollars a dozen. To compensate for this scarcity of primate eggs, those of the laboratory animals are fairly easy to collect, and the domestic pig is a prime source. I have myself handled 2,500 or more sow's eggs and used to demonstrate them annually to my medical students. Any high-school biology teacher who lives near a slaughterhouse, if he will learn the tricks of finding and handling them, can show his boys and girls this striking evidence of the unity of living things.

The clear outer membrane of the mammalian egg is tough, like a very stiff gelatine solution, stiffer yet than a housewife would serve for dessert. I have often pushed the eggs of rabbits and pigs over the bottom of a dish of salt solution, using a coarse needle which under the microscope looked like a poker pushing a grape. The tough egg membrane easily

PLATE VII. Structure of the ovary. *A*, diagram of a section through ovary illustrating the structures described in the text. From an article by the author in *Physiological Reviews*, by permission of the editor. *B*, photograph of small part of a microscopical section of a monkey's ovary. The letter *c* indicates the cortex of the ovary, containing egg cells not yet surrounded by follicles. *a.f.*, an atretic (degenerating) follicle. Photograph magnified 50 diameters. *C*, photograph of section of ripe egg of Rhesus monkey, in its hillock inside a large follicle (Corner collection, no. 100). Magnified 100 diameters. *D*, photograph of living human egg, recovered from Fallopian tube at operation. Preparation by Warren H. Lewis (Carnegie collection, no. 6289). Magnified 200 times.

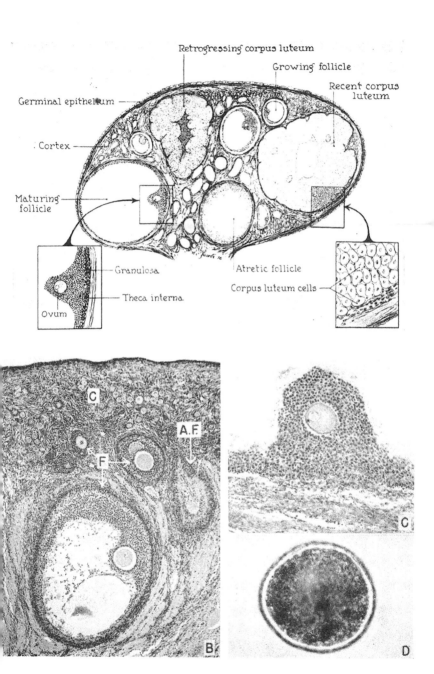

Retrogressing corpus luteum

Growing follicle

Recent corpus luteum

Germinal epithelium

Cortex

Maturing follicle

Granulosa

Theca interna

Ovum

Atretic follicle

Corpus luteum cells

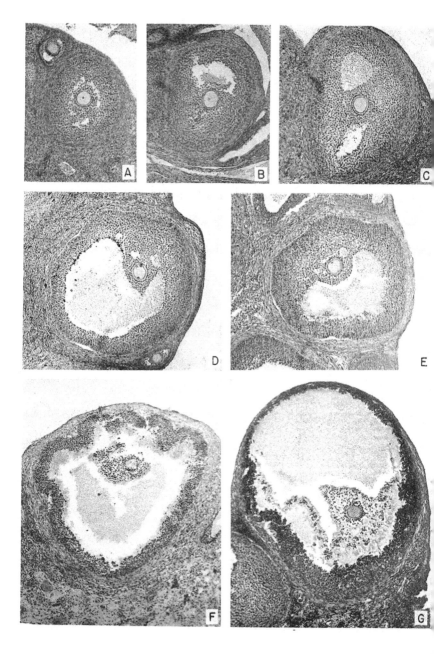

resists such handling, but if a thin glass microscopic cover slip is allowed to settle down upon an egg in a drop of water, the clear membrane splits open under its pressure like the skin of a grape and pops out its soft contents. It seems to me quite remarkable that the infinitesimal sperm cell can push its way through this barrier.

Some eggs, like those of the pig, are heavily laden with fat globules (*yolk*); others, e.g. those of the rabbit, are so clear that the nucleus, inevitable part of a living cell, can be descried in the fresh egg. The human egg, as will be seen from our photograph (Plate VII, *D*), is moderately filled with yolk granules and the nucleus cannot be seen. If the microscopist wishes to study details of the nucleus in any species he must "fix" the egg, i.e. kill and harden it with chemicals, and stain the nucleus with a dye solution.

However different the eggs of birds and of mammals may seem at first sight, they are fundamentally alike. Each is a single cell, with a nucleus no bigger than that of many of the ordinary cells of the body. The fact that the bird's egg is very large for a single cell, and that even the mammal's egg is the largest cell in the mammalian body, is due to the inclusion of a considerable amount of stored food substance in the cell. In the hen this yolk is enough to feed the chick until it hatches; in mammals it is only a few grains of fat and protein, sufficient to provide energy for growth and cell division for a few days, until the fertilized egg reaches the uterus.

Since the bird's egg needs protection from harsh external conditions—sunlight, dryness, a rough nest—it is provided with a hard shell secreted about it by the oviduct after it leaves the ovary. The mammalian egg, which is destined not

PLATE VIII. Stages in the development of the Graafian follicle of the rat. Note the gradual enlargement of the cavity. In *G* the cells holding the egg to the wall of the follicle have begun to degenerate, and the follicle is ready to rupture. Magnified 60 times. From the *Anatomical Record*, by courtesy of J. L. Boling and the Wistar Institute of Anatomy and Biology.

to leave a soft, moist, dark environment within its mother until it is ready for birth, has no shell at all. The clear zone that surrounds it is comparable with the shell membrane of the hen's egg, familiar to everyone who has peeled the shell off a boiled egg. The bird's egg also receives in the oviduct, before the shell is laid on, a layer of albumen ("white of egg") which no doubt helps to cushion the yolk and has some nutritive value for the growing embryonic bird, but is chiefly important because of its property of holding water and thus preventing the egg from drying out by evaporation through the shell. This is another protection the mammalian egg does not require. It is therefore much smaller than the bird's egg, because, in the first place, it lacks these massive provisions for independent existence; but also for a positive reason which my mathematically minded readers may have perceived. The mammalian egg, after only a few days of total dependence upon its paltry yolk, gets its nourishment by absorption of food and water by diffusion through its surface from the uterine fluid in which it reposes. How much it gets depends upon the area its surface presents to its surroundings. How much it needs depends upon its volume. Geometry teaches that as dimensions increase, surfaces increase as the square of the radius, but volumes increase as the cube. As organic bodies increase in size, therefore, the ratio between volume and surface becomes less favorable. This rule, which has been invoked to explain why animals do not grow to limitless dimensions, probably operates also to keep all sorts of cells, including eggs, within effective limits of size.

The fate of unfertilized eggs. Not all the eggs formed in the ovary, nor even a large proportion of them, go on to reach fruition. Most domestic mammals shed eggs from the ovary at regular intervals, the human one egg per month approximately, the guinea pig four or five eggs every fifteen days, the sow an average of about a dozen every twenty-one days. If there is no mating in a given cycle, those eggs proceed to degenerate while in the oviduct. Others degenerate in the ovary, sometimes even before well-developed follicles have formed about them, sometimes in large follicles. When this happens, the follicle also ultimately degenerates and disappears from the ovary. Such follicles are shown in Plate VII, *A, B.*

A Swedish investigator, Haggström, who counted the eggs in both ovaries of a 22-year-old woman, found about 420,000. Yet a woman who sheds one egg per month without interruption by pregnancy or illness during her entire life, from say the 12th to the 48th year, cannot use up more than 430-odd eggs. The most prolific egg producer among mammals, the sow, might possibly shed a total of 3,000 to 3,500 eggs, allowing ten years of ovarian activity not interrupted by pregnancy, and assuming the very high average of 20 eggs at each 3-weekly cycle; but she has vastly more than this in the ovaries at birth. Whether or not there is new formation of eggs during adult life (as we discussed above), there is evidently a large overproduction of eggs in the ovaries. This, and a corresponding but enormously greater overproduction of sperm cells, is thought to be a survival from earlier evolutionary stages when germ cells were discharged into the water to take the risk of enemies and mischance.

When the reproductive period of life is over, there are still many eggs in the ovaries. These gradually diminish in number, but some of them may persist to advanced ages.

The corpus luteum. When the eggs of a frog or fish are spawned into the water the ovary has done its work. There is nothing more it can do for the eggs. It shrinks back to the insignificant bulk it had before the eggs began to ripen and nothing more is heard from it until the next breeding season. Not so in the mammals. Their eggs are not thrust into the outside world. The mother is still intimately responsible for them; they must be nourished and sheltered within her for weeks or months to come. The uterus is to be altered to receive them and keep them while they develop, the mammary gland must be signaled to grow and prepare milk for them when they are born. The ovary still has ahead of it the task of getting these things done.

The Graafian follicle therefore does not shrivel away after shedding the egg. Indeed, it has scarcely had time to collapse

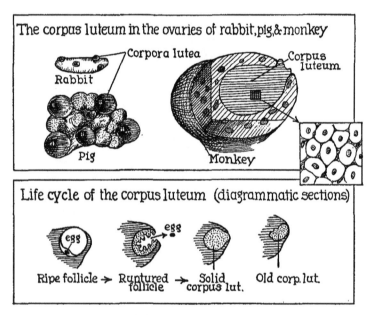

FIG. 9. Diagram illustrating the structure and history of the corpus luteum.

before it is being transformed into a *corpus luteum* (Fig. 9) and begins to function as an organ of internal secretion for the benefit of the embryo. The accompanying illustrations (Plate IX) give an idea of the situation and appearance of the corpus luteum. This remarkable change is brought about by growth of the cells that lined the cavity, which become so much larger that the inner wall of the follicle, folded by its collapse, becomes thick and firm and converts itself into a glandular body occupying the site of the follicle (Plate IX, *A, B*). As the lining grows thicker, blood vessels creep in from the surrounding part of the ovary and make a network that carries blood past every one of the large cells (Plate IX, *C*). These cells become laden with a peculiar kind of fatty material, and in animals whose fat is yellowish, for example the cow, the transformed follicles are bright yellow in hue, becoming indeed just about the most brilliantly colored objects in the whole body. For this reason they were long ago named *corpora lutea*, yellow bodies; but in animals whose body fat is white, as the sow, sheep, rat, and rabbit, they appear pink or whitish. The human corpus luteum forms a mass almost three-quarters of an inch in diameter, with a folded wall, bright orange in color, about a grayish core of fibrous tissue.

Animals which shed one egg at a time have, of course, only one corpus luteum in each cycle of the ovary; animals which bear multiple litters have a corpus luteum for each egg, i.e. for each follicle that ruptured. The sow averages ten corpora lutea in a batch, and may have twenty-five, which is the number of the largest litter of pigs ever recorded. The rat can have eighteen, the guinea pig two to four, dogs of various breeds as many corpora lutea as there are puppies in the litter of the breed.[3]

These yellow bodies of the ovary have been puzzling to scientists ever since they were first described by Regner de

[3] There are interesting exceptions to the statements in the foregoing paragraph. In the first place, human females occasionally shed two eggs at one time; if both are fertilized twins will be produced. In the case of identical twins there is only one egg, which forms two infants. Triplets usually come from two eggs, one of which gives twins. In the case of the Dionne quintuplets it is conjectured from indirect evidence (i.e. the close resemblance of the 5 sisters) that they all came from one egg. In animals with multiple litters it is often the case that not all the fertilized eggs develop successfully; then obviously there will be more corpora lutea in the ovary than infants in the litter. On the other hand it is possible, though uncommon, to have more infants than corpora lutea, for one follicle may contain two eggs (a rare event) or one or more of the eggs may develop into single-ovum twins.

Graaf in 1672. A French medical student who wrote a thesis about them in 1909 listed twenty-five different incorrect hypotheses about their function; but already in 1898 Louis-Auguste Prenant had suggested that they might be glands of internal secretion, making some sort of hormone for the benefit of the eggs with which they are associated. Now that we know more about such glands, any microscopist can see that the corpora lutea have the signs of endocrine function written all over them. The large, imposing cells, built into a mass that communicates with the rest of the body only by the blood vessels; the delicate texture, scarcely supported by connective tissue; the wealth of blood supply that reaches every cell—these are the telltale evidences that the corpora lutea are indeed organs of internal secretion, and that whatever product they secrete must be poured into the blood and carried away to exert its effect upon some other organ. The full story of the corpus luteum hormone, as we know it now, will be told in Chapter V.

The life of the corpus luteum is relatively short. If the egg is fertilized, the corresponding corpus luteum persists through the greater part, if not all, of pregnancy. If the egg is not fertilized, the corpus luteum has an active life of only about two weeks before it begins to degenerate. In the human cycle of four weeks a fresh corpus luteum is present, therefore, about half the time. The older corpora are visible in various stages of degeneration. Five or six months after the formation of a corpus luteum all traces of it have disappeared.

The oviducts. When an egg is discharged from the ovary,

PLATE IX. The corpus luteum of the Rhesus monkey. *A*, ovary split into two parts and laid open to show the corpus luteum. Magnified 4 times. Courtesy of C. G. Hartman. *B*, section through ovary showing a large corpus luteum (*Corner collection, no. 187*). Magnified 10 times. *C*, small part of corpus luteum, magnified 250 times to show the cells. The narrow clear spaces between the cells, bordered by small dark nuclei, are capillary blood vessels. At the left 6 cells and parts of a blood capillary have been outlined with ink to show how each cell is in contact with a blood vessel.

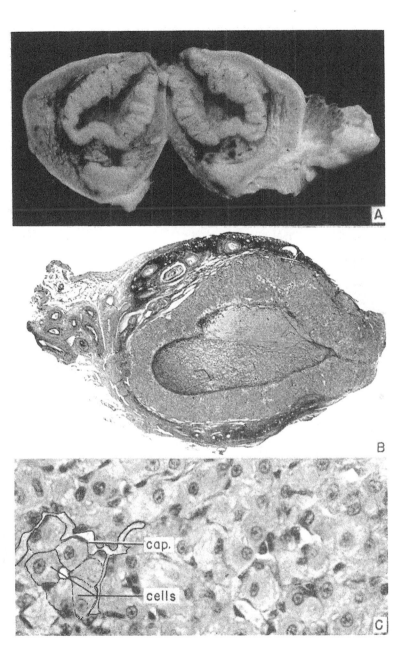

cap.

cells

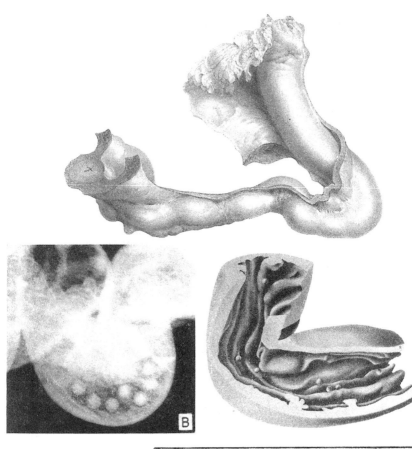

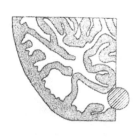

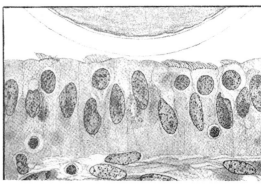

it is received into the open end of one of the two *oviducts*, tubular conduits, each (in the human species) about 11.5 centimeters (4½ inches) long (Plates V and X). Medical men and the general public usually call them "Fallopian tubes," although Gabriele Fallopio (1523-1562) was not the first to mention them and had no idea of their real function; he thought they were ventilators to let noxious vapors out of the uterus. The walls of the oviducts are made, like the intestines, of involuntary muscle cells. Their lining is a velvety membrane which follows their channel all the way to the uterus and joins the lining of that organ. The cells on the surface of this membrane are beset with fine hairlike processes ("cilia") lashing continuously downward, and thus producing a current through the oviducts toward the uterus. These cilia may be seen in Plate X, *E*. At their free ends, near the ovaries, the oviducts open directly into the abdominal cavity by handsome trumpet-shaped expansions with fringed edges covered by the velvety red lining tissue. One of the fringes of each oviduct runs right on to the ovary.

When we say the oviduct opens into the abdominal cavity, we must not forget that the "cavity" is actually packed full of intestines. When an egg escapes from the ovary, it does not pop into a large vacant space; it merely glides in a thin film

PLATE X. *A* (at top), oviduct (Fallopian tube) of Rhesus monkey, drawn y J. F. Didusch from preparation by author. Enlarged 4 times. *B*, photograph f living eggs of a mouse, in passage through the oviduct. The eggs are seen hrough the walls of the oviduct, which is exceedingly thin in this small 1ammal. Magnified about 45 times. Courtesy of H. O. Burdick. *C*, model of a •art of the oviduct of a rat, showing eggs in passage. Magnified about 33 times. 'rom an article by G. C. Huber, by courtesy of the Wistar Institute of 1natomy and Biology. *D*, diagram showing comparative size of the egg of the abbit and the folds of the lining of the oviduct. Courtesy of G. H. Parker. *E*, omparative size of the human egg and the cilia of the lining cells of the viduct. The cilia are seen as little brushlike clumps on the free ends of some f the tall cells. Enlarged 600 times. This drawing was made by combining part f a human egg described by Warren H. Lewis (see Plate VII, *D*) with a •icture of the epithelium of the oviduct from a paper by F. F. Snyder in the *Bulletin of Johns Hopkins Hospital.*

of moisture between the smooth surfaces of the organs in the region of the ovary. The open funnel of the oviduct is directly at hand, and moreover the moisture in which the egg drifts is constantly drawn into the oviducts, carrying the egg with it, by action of the cilia mentioned above. Small particles of carmine or even foreign eggs, introduced into the lower abdominal cavity by the experimenter, are within a few hours carried down the oviducts toward the uterus.

It is even possible for the egg to drift across from one ovary to the opposite oviduct, a distance of roughly 3 or 4 centimeters (1 to 2 inches), and therefore a woman who has had one ovary and the opposite Fallopian tube removed is not necessarily sterile. In some animals, e.g. the sow, the oviduct expands into a voluminous sac partly enclosing the ovary; in the dog and cat the enclosure is almost complete; in the rat and mouse it is quite complete and eggs are obliged to travel down the oviduct corresponding to the ovary from which they came.

How are the eggs transported? We know that this trumpet-like capsular part of the oviduct, just mentioned, throws itself during life into squirming movements which are especially active at the time the eggs are discharged. This may help draw the eggs into the oviduct. How they are pushed along toward the uterus, once they are in the tubular canal of the oviduct, is at present under discussion. When there are several eggs (that is, in animals which bear several young at a time) the eggs travel together at first, sticking together in a little web of cellular debris they have brought with them from the ovary, but after a few hours this entanglement dissolves and the eggs travel free and bare, though still more or less closely together. As will be seen from Plate X, *C, D,* the lining of the tubes forms voluminous folds, so that the available space is hardly larger than necessary to permit passage of the eggs. It used to be thought without question that the eggs are brushed along by action of the lashlike cilia of the surface cells. This is still not out of the question, in spite of the relatively small size of the cilia as compared to the eggs—about in the proportion of an eyelash to an orange (Plate X, *E*). With a bunch of eyelashes or something similar, for example a tiny camel's hair brush, it does not take much effort to roll along an orange floating in water, as the eggs float in the fluid contents of the oviduct.

This supposition, however, has its weak points. In the first place there are animals in which the oviduct is not provided with cilia throughout its entire length. In the second place, there is a remarkable fact which cannot easily be explained on the basis of ciliary transport, namely that the oviducts of different species of animals are of very different lengths, and yet with only a few known exceptions, the eggs make their journey through them in about the same time, reaching the uterus in 3 to 3½ days. The oviduct of the sow is about forty times as long as that of the mouse, therefore the eggs must travel forty times as fast. The cilia, however, certainly do not beat that much more rapidly.

What is more, the cilia beat with more or less uniform motion, while the eggs do not travel at uniform speed. A former colleague of mine, Dr. Dorothy Andersen, once collected at a packing house a very large number of oviducts of swine containing eggs. She cut up each one into 5 segments and examined each segment separately to see whether it contained eggs. She found that it is common to find eggs in the middle segments, but rare to find them in the first and in the last parts of the tube—in other words, the eggs are rushed through the first fifth, transported very slowly through the middle stretch, and then hurried through the last part into the uterus. A similar and even more accurate observation has since been made in the mouse (W. H. Lewis and Wright).

All these difficulties lead us to suppose that the eggs are really transported by contractions of the muscle fibers in the walls of the oviducts, which move them along by a "milking" action. Such a mechanism is common in the body. That is how food is shoved along in the intestines. That is how a horse gets water through his esophagus up to his stomach when his mouth is away down in the pond. Similar contractions of the walls of the ureter force urine from kidneys to bladder, no matter what position the body may be in with respect to gravity. We know that the muscular walls of the oviducts undergo contractions which could move the eggs, and we know also that these contractions are under the influence of the hormones of the ovary, changing their rhythm and intensity at the very time the eggs are in transport. Burdick, Pincus, and Whitney have been able to lock the ova in the oviducts by administration of an ovarian hormone. Most students of this problem now think, therefore, that the chief method of transport of the eggs is by rhythmic contractions of the tubal muscles, and that the cilia play at most only a secondary role.

I have long thought that we ought not to emphasize the oviduct solely as an organ for transporting the ova, but rather as a means of delaying their transportation. We are going to see (in Chapter V) that the mammalian embryo, reaching the uterus naked, delicate, and yolkless on the fourth day after leaving the ovary, requires immediate nourishment and a soft succulent place in which to grow. The uterus must have time to get ready for its exigent tenant. If the embryos arrive too early they cannot develop. I believe that one of the most important functions of the oviduct is to hold back the eggs until the uterus is ready for them.

The uterus. When the eggs pass from the oviduct into the uterus they find themselves in a chamber of larger size, with heavier and more muscular walls.

The uterus is built on fundamentally the same plan in all mammals, although its form varies a great deal in different species. It consists basically of two tubular canals, one right and one left, corresponding to the two ovaries and oviducts.

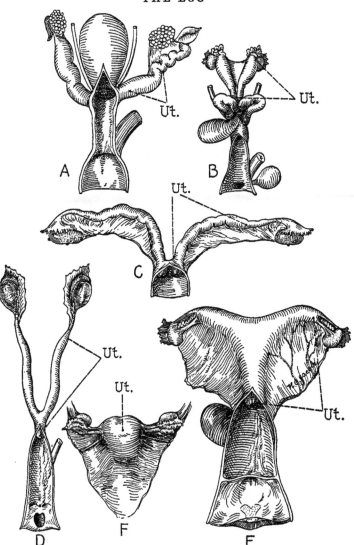

Fig. 10. Form of the uterus in a series of mammals, illustrating the various degrees to which the two horns of the uterus are separate or fused. *A*, monotreme (Echidna); *B*, marsupial (opossum); *C*, rodent (rabbit); *D*, carnivore (dog); *E*, ungulate (mare); *F*, primate (Rhesus monkey). From *Physiology of the Uterus, with Clinical Correlations*, by S. R. M. Reynolds, by courtesy of the author and Paul B. Hoeber, Inc.

Each canal is really the continuation of one of the oviducts. At their lower ends the uterus enters the last part of the genital tract, namely the *vagina*. In most animals the two canals of the uterus unite before they enter the vagina, forming a Y-shaped organ, with a single stem and two horns

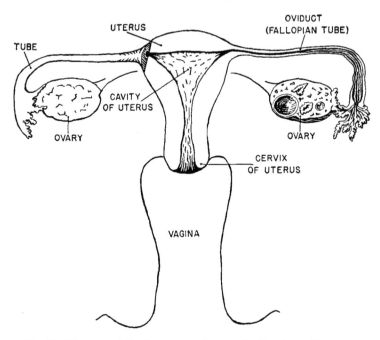

Fig. 11. Diagram of the human female reproductive tract. The uterus, and the oviduct at the right, are depicted as if laid open by removing the nearer half. The vagina is drawn as if fully distended. In part after a drawing by R. L. Dickinson in his *Human Sex Anatomy*.

(Fig. 10, *E*), but the extent of this fusion differs very much in different animals. In rabbits the two horns remain entirely separate and enter the vagina independently, side by side (Fig. 10, *C*). In monkeys, apes, and humans the opposite extreme is found, for even the horns fuse together, closing the Y and making a single-chambered uterus into which the two oviducts are inserted as shown in the diagram, Fig 11.

Even in these animals, however, the uterus is double in its embryonic development. In Rhesus monkeys (Fig. 10, *F*) and in human infants there is a little notch at the top of the organ, marking the last trace of the doubling. In the other species shown in Fig. 10, namely echidna (*A*), opossum (*B*), dog (*D*), and mare (*E*) we find various degrees of fusion of the two horns. Sometimes in women the process of fusion fails to be completed, leaving a bicornuate (two-horned) uterus which under certain circumstances may puzzle the gynecologists or even be mistaken for a tumor. In a general way it may be said that animals which bear a single infant, or twins, at one time, have one-chambered uteri or short uterine horns, while those that bear multiple litters have well-separated horns. I hardly know which of these types offers the most striking picture, late in pregnancy when the uterus reaches its largest dimensions—the human, for example, with its infant ensconced in a single huge chamber, the cow with an unborn calf in one enormous sac with a little empty horn beside it, or a sow with two long uterine horns each distended like two great strings of sausage, with five to eight 12-inch pigs in each link.

The lining of the uterus. The longest act of the drama of reproduction is played in the cavity of the uterus. From the first week after the egg is shed from the ovary, until the day of birth, the infant knows no other environment, and depends absolutely upon the reactions of growth and chemical exchange that take place in these walls that shut it off from everything else in the world. Here occur, as we shall see, some of the most remarkable and important interactions of the hormones of the ovary and the tissues that guard the embryo, and here is the seat of the process of menstruation, strange phenomenon that is an outward sign of human participation in the cosmic tides.

Because the inner layer or lining of the uterus and what goes on in it will occupy much of this book, we may as well introduce its technical name at this point, to save printing two words for one every time it is mentioned: *endometrium*, from Greek *endo*, within, and *metron*, the uterus. It is a layer about 5 millimeters (1/5 inch) thick, lining the cavity and therefore applied to the inner surface of the pear-shaped muscular parts of the organ. At the upper end of the uterus it blends with the lining of the oviducts as they enter, at the lower end it continues on to become the lining of the vagina. It looks rather like pink or red velvet, slightly moistened. In most animals it is thrown into delicate folds, but in the human uterus it is relatively smooth. Upon the cells and secretions of this layer the embryo is to depend for everything it needs during gestation.

It is always difficult to convey in nontechnical terms an idea of the finer structure of the tissues of the body. In a book for general readers, microscopical anatomy is like mathematics in books on astronomy or physics—something to be avoided if possible. Yet physiology without cell structure means less than Einstein without calculus. Therefore let us buckle down together for a few pages and try to build up a picture of the cell structure of the endometrium for subsequent use.

The effort would be much easier if we could sit down together in my laboratory and prepare a specimen as shown in Fig. 12. Taking a preserved human uterus from a jar of formalin, we cut it in two lengthwise with a sharp knife so that we can look into the cavity (Fig. 12, *A*). Then we cut out a horizontal slab of uterine tissue (*B*) and from this we detach a little block running down through the endometrium into the muscle (*C*). This we shall place on the table, so that its upper side will be that which formed the surface of the lining, facing the cavity; i.e. like a cube of melon with the rind downward and the pulp upward (Fig. 12, *C*). After we have studied and sketched it under low magnification, we shall cut off a very thin slice (technically

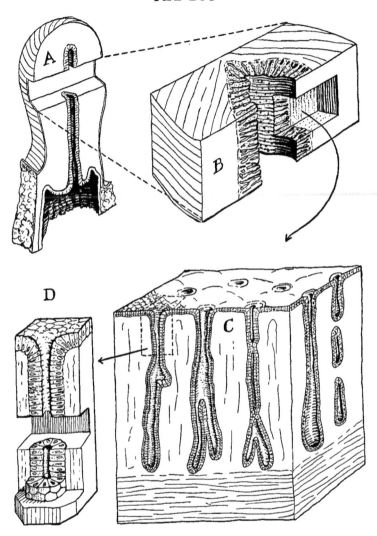

FIG. 12. Block diagram showing construction of the lining of the uterus (endometrium). At *A* the uterus is represented as if cut in two lengthwise, to show its lining. At *B* is shown a block cut from the uterus; a small part of this is represented at *C*, turned so that the inner surface of the endometrium is upward, showing the glands. At *D* a small part of *C* is drawn still more enlarged, to show that the glands are really cell-lined tubes dipping down from the surface epithelium.

section) from one side, stain it with appropriate dyes and photograph it through the microscope (Plates XVII, XXI). When studying the two-horned uteri of small animals such as the rat, rabbit, or guinea pig, we usually cut our blocks from the whole thickness of the tubular horns, as one slices a banana, and therefore the thin sections for microscopic study are round, with the uterine cavity showing in the center (see Plate XVII, *B*, in comparison with *A* of the same plate).

We find that the surface is paved with a single layer of tall cells, and that at frequent intervals this surface layer pushes down into the depth of the endometrium, forming fingerlike tubes, closed at the end, which reach almost to the muscle (Fig. 12, *C, D*). These tubes are supported by spongy connective tissue, and between them there is a network of capillary blood vessels supplied by arteries. They are in fact actually glands, able as shown in Fig. 13, to take water and the "makings" of nutritive substances from the blood vessels, build them up into foodstuffs for the early embryo, and discharge the resultant secretion into the cavity of the uterus. The endometrium is therefore something like a quick-lunch counter, with a supply of raw foods in the rear (in the blood stream), a row of cooks and waiters (the gland cells) and a line of customers (the cells of the embryo). The outfit does not however function in this way all the time; it secretes nutritive materials, practically speaking, only when an egg is likely to be present. How all this is regulated by the hormones of the ovary will be explained in Chapter V. To paraphrase a saying of Robert Boyle, the endometrium looks like so much velvet, yet there are strange things performed in it.

The cervix and vagina. The lower end of the uterus projects downward into the vagina as shown in Fig. 11. This part of the organ is known as the neck or *cervix*. Its lining is full of glands which secrete mucus.

The lowest part of the genital canal, the vagina, is lined with a membrane made of cells many layers thick (Plate XIII, *A*), closely resembling the structure of skin, except that the latter is dry and somewhat scaly, while the vaginal

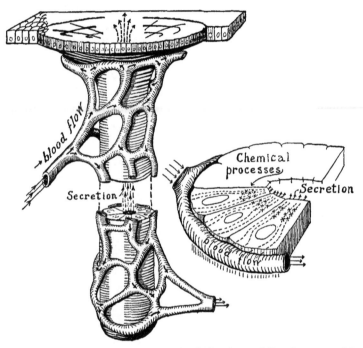

FIG. 13. Diagram representing a gland like those of the uterus, consisting of a tube of cells dipping down from the surface. This is surrounded by a network of capillary blood vessels, from which water and other substances pass through the gland cells, undergoing chemical elaboration, and are discharged into the central channel of the gland and thus reach the cavity of the uterus.

lining is moist. The lining of the vagina, in fact, becomes continuous with the skin of the outside of the body, at the vaginal orifice, just as the membranes of the nose, mouth, and lower intestine become continuous with the skin.

Fertilization and segmentation of the egg. Sperm cells deposited in the vagina by the male make their way up through the canal of the cervix and body of the uterus and into the oviduct. This journey is accomplished in a few hours, so that the descending egg and the ascending sperm cells meet in the Fallopian tube. There the process of fertilization takes place by entry of one sperm cell into the egg, as described for the sea urchin in the last chapter.

The egg now begins to divide. The process of division has not yet been observed in the human egg, but it has been well studied in many animals by the drastic and expensive method of killing animals at successive stages, a few hours apart, after mating so that the eggs can be found and studied under the microscope. In recent years the dividing eggs of rats, mice, rabbits, and monkeys have, through the skill of Warren H. Lewis of the Carnegie Embryological Laboratory and his various associates, been successfully removed, kept in dishes of salt solution at body temperature, and photographed in motion pictures. Our illustration (Plate XI) is taken from an excellent series of still photographs of the rabbit egg taken by P. W. Gregory in the same laboratory. To those who have not seen such pictures, this series will cause surprise chiefly by its resemblance to the dividing sea urchin eggs of Dr. Ethel Browne Harvey's series (Plate IV).

In most mammals the embryos pass from the oviduct to the uterus late on the third day or on the fourth. There is some reason to think the same is true in the human. By this time the embryos are at least in the four-cell stage and in some animals (e.g. the rabbit) they have divided even more fully, entering the uterus as little clumps of cells called *morulae* from the Latin word for mulberry, which they resemble. The

PLATE XI. Division of the fertilized egg of the rabbit. *A*, one cell sta[ge] *B*, two cell stage, 25 hours after mating. *C*, four cells. *D* to *G*, 4 to 32 cel[ls] *H*, morula stage (solid mass of cells). *I*, first signs of hollowing. *J*, *K*, holl[ow] stages (blastocysts) 90 and 92 hours after mating. Magnified about 140 tim[es] From *Contributions to Embryology*, Carnegie Institution of Washington, courtesy of P. W. Gregory.

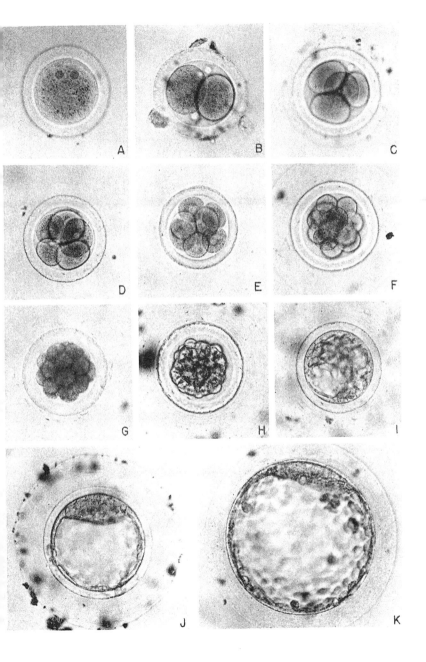

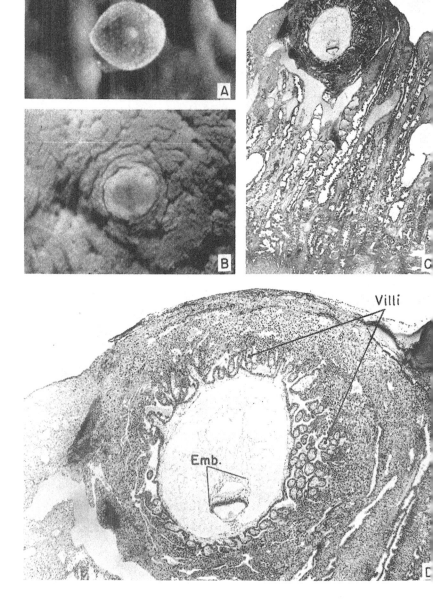

Villi

Emb.

embryo soon becomes a hollow sphere, with a little mass of cells at one side (Plate XI, *J* and *K*). This *inner cell mass* is to become the embryo proper; the remainder of the spherical cyst becomes the embryonic membranes.

During the first few days after arrival in the uterus the embryos are free and unattached. In animals bearing only one infant at a time, as usual in the human species, the embryo simply settles down somewhere on the inner wall of the uterine cavity. Species bearing several young have long uterine horns to provide room for them all. In such animals the free embryos must be moved along the uterus, by a kind of squirming movement of the uterine walls, until they are spaced at regular intervals.

Implantation. Attachment or implantation begins, in most species, about the 7th day, but in some (e.g. the sow) as late as the 13th. The nature and extent of the attachment of infant to mother, though fundamentally the same in all mammals, varies a good deal, in its structural character, in the various orders of the Mammalia. In the ungulates (hoofed animals), for instance, the attachment is not very intimate. In the sow the voluminous membranes of the embryos are simply apposed to the inside of the uterus, like a string of crumpled bags fitted inside the long uterine horns, and the embryos get their nutritive fluids and the oxygen they breathe,

PLATE XII. Implantation of the embryo in the primate uterus. *A*, embryo of Rhesus monkey (Carnegie C. 610) in the blastocyst stage, 9 days old, just settling down on the lining of the uterus. The little white spot in the center is the "inner cell mass," destined to become the embryo proper. Magnified 50 times. *B*, human embryo about 12 days old (Carnegie 7700) which has burrowed into the uterine lining. Magnified 12 times. *C*, section of a very similar human embryo (Carnegie 7802) showing how it lies within the endometrium. The glands of the uterus are in a state of progestational proliferation under the influence of the corpus luteum hormone (see Chapter V). Magnified 10 times. *D*, portion of same section magnified 80 times, to show the embryo proper (*emb.*) and the placental villi, which are beginning to grow out from the envelope (chorion) of the embryo. *A* by courtesy of G. L. Streeter, C. H. Heuser, and C. G. Hartman; *B*, *C*, and *D*, by courtesy of A. T. Hertig and John Rock. Photographs by Chester Reather.

by filtration through these apposed membranes. When the pigs are born the membranes ("afterbirth") simply peel off the lining of the uterus leaving the latter more or less intact.

In most mammals, however, the contact becomes much more intimate. The membranes send down long rootlike processes

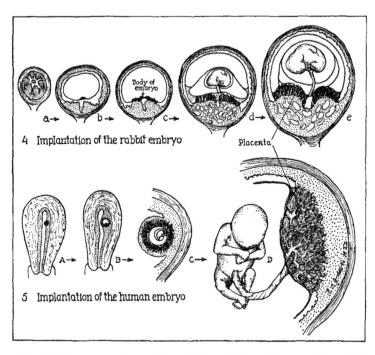

4 Implantation of the rabbit embryo

Placenta

5 Implantation of the human embryo

FIG. 14. Diagrams showing implantation of the embryo, in the rabbit and human. In the rabbit figures the uterus is represented as if cut transversely, as one slices a banana; in the human figures the uterus is cut lengthwise as one halves a pear. From *Attaining Manhood*, by George W. Corner, by courtesy of Harper and Brothers.

into the uterine lining or endometrium. This is illustrated in the diagram of the rabbit's implantation (Fig. 14). In the human and a few other animals, the early embryo settles down not merely on the inner surface of the uterus, but actually burrows down into the lining, and sends out its rootlike processes (villi) all about it. As it grows it bulges into the uterine cavity, remaining rooted at its base, and thus forms a definite organ of attachment, the *placenta*. The sketch, Fig. 14, gives a diagrammatic idea of this arrangement, and Plate XII illustrates some of the details of implantation in man and monkey, from the unique specimens of the Carnegie Embryological Collection.

Within the placenta, blood vessels of the embryo ramify in close proximity to the blood stream of the mother. Nutritive substances and oxygen filter from the uterus through these blood vessels into the infant's blood; waste substances and carbon dioxide filter out. The infant, completely immersed in the fluid contents of its dark chamber, thus gets only such substances as can be brought to it dissolved in the mother's blood and filtered through the placenta.

Preparation of the uterus for implantation. It may make the next chapters clearer if we anticipate slightly at this point our discussion of the corpus luteum hormone (Chapter V). While the embryo is floating into the uterus prior to its attachment, it requires nourishment from the uterus. To provide this, and also to favor the invasion of the maternal tissues by the placenta, the hormone made by the corpus luteum (progesterone), acts on the uterus, causing great activity and growth of its tubular glands. Without this preparation the embryo, arriving in the uterus, would be unable to develop, like the seed told of in the Biblical parable, that fell on stony ground.

THE OVARY AS TIMEPIECE

"*They bounded to the horizon's edge
And searched with the sun's privilege*

.   .   .   .

*Saw the endless wrack of the firmament
And the sailing moon where the cloud was rent,
And through man and woman and sea and star
Saw the dance of Nature forward and far,
Through worlds and races and terms and times
Saw musical order and pairing rhymes.*"

RALPH WALDO EMERSON, *The Poet.*

⊷§ CHAPTER III §⊶

THE OVARY AS TIMEPIECE

OUR western plainsmen used to watch, in August and September, milling herds of bison, blackening the prairie for miles. It was no uncommon thing to see thousands of them, eddying and wheeling about under a dense cloud of dust raised by the bulls as they pawed in the dirt or engaged in desperate combat. In these herds the males were continually following the females and mating with them. The whole mass was in constant motion, all bellowing at once in deep and hollow sounds, which mingling together seemed at the distance of a mile or two like the noise of distant thunder.[1] This was the yearly period of estrus,[2] the mating time, when the females were ready to produce their eggs, and the

[1] This passage is largely a quotation from George Catlin, *The North American Indians*, vol. 1 (p. 280 in the Edinburgh reprint of 1926). Catlin's rather discreet painting of such a scene occurs in the same volume, fig. 105.

[2] *Estrus* is the technical term for recurrent periods of sexual excitement in animals, popularly called "heat." It was introduced in 1901 by Walter Heape, a prominent student of the physiology of reproduction. The word comes from the insect described by Virgil:

"About the groves of Silarus and Alburnus evergreen
In holm-oak swarms an insect
We call the gadfly ('oestrus' is the Greek name for it)—
A brute with a shrill buzz that drives whole herds crazy
Scattering through the woods, till sky and woods and the banks of
Bone-dry rivers are stunned and go mad with their bellowing."
(Georgics, Book III, C. Day Lewis's translation.)

In its Latin neuter form *oestrum* this word long ago became an English word meaning any recurrent excitement, e.g. the poetic frenzy. Heape adopted the masculine form as his special technical term. In England it is spelled *oestrus* and the first syllable is pronounced ĕ as in "me." In the United States, following a general trend of our language, it is now commonly spelled estrus and pronounced with ĕ as in "west." The adjectival form is *estrous*; cf. *mucus, mucous*. The interval between two estrous periods, in Heape's terminology, is *diestrus;* a long period between sexual seasons (as for example in sheep during the winter) is *anestrus*.

males to fertilize them. Such a season of mating is well-nigh universal in nature, though fortunately not all the denizens of earth react as violently as the bison. The rhythm of sex is manifested in infinite variety, assuming every aspect; now to our human eyes tensely dramatic, now gently romantic, now bestial or merely matter-of-fact, sometimes even comical.

In springtime two robins nest beneath my window, and soon a group of eggs in the nest gives evidence that the ovaries of the female bird have responded to the rhythm of the year. Twice a year my neighbor's spaniel gives unmistakable signs that she is ready to mate, and I know, being an anatomist, that if we could inspect her ovaries we should see the Graafian follicles enlarging, and the microscope would show us a crop of ripening eggs.

The ovary as a timepiece runs at curiously different rates in different animals. Many creatures living wild, both plants and animals, necessarily time their breeding with the seasons of the year, because their offspring must begin life when conditions of temperature, shelter, light and food are most favorable. Hence the vernal growth of plants and all the annual breeding seasons of animals such as those of migratory birds and of the fish, salmon and shad for example, that swarm into the bays and rivers every spring, teeming with roe and milt and seeking a sheltered place in which to spawn. Many marine plants and animals have reproductive cycles controlled by the tides and therefore breed at intervals of a month or multiples of a month. There are seaweeds, for instance, that fruit only on the highest tides, and worms that breed at particular phases of the moon. The Japanese palolo, an annelid worm living on the sea bottom, swarms to the surface to breed on four nights of every year, namely on the new moon and the full moon of October and November (Appendix II, note 2).

In the course of evolution, however, many animals have adopted cycles not directly related to the yearly seasons or the tides. Some of these are domesticated species which man has freed from dependence upon the wild crops and has improved for more rapid breeding; others have acquired their cycles for no obvious reason. The shortest cycle is that of the domestic fowl, which lays an egg once a day; the longest, that of the locusts that come swarming from the ground at intervals of seventeen years, in obedience to some obscure signal, to deposit their eggs and then to die.

In mammals, and in the human race, the ovary is no less cyclical than in lower animals. Many wild mammals have an annual cycle so timed that they may bring forth their infants when food is plentiful for the mothers. Other species have estrous cycles at intervals throughout the year, or a sexual season of several cycles at a favorable time of the year. Rats and mice have very short cycles, ovulating every 4½ to 5 days, except when the cycles are interrupted by pregnancy. The guinea pig has a 15-day cycle. Cows, mares, and swine have estrous periods at 21-day intervals throughout the year. Sheep have several cycles in the late summer; during the winter they are anestrous. Dogs and cats have two or three estrous periods each year; so, apparently, have lions. Many other carnivores are monestrous (having one period each year). Not only do the time intervals vary in different species, but there are all sorts of different behavior patterns at estrus and a good many differences in the details of internal physiology.

The cycle of the sow. To make this matter clear, let us follow through the cycle of one particular species and then discuss some of the variations. That valuable creature, the domestic sow, will serve us admirably for this purpose. She has an estrous cycle of 3 weeks' duration. During 2½ weeks of each cyclic interval she goes about her usual program of eating and sleeping and if there is a boar in the herd she shows no interest in him. Then there is a change of mood and behavior. For 3 days she undergoes a well-marked phase of estrous excitability, marked by restlessness, diminished appetite, and heightened sexual interest. If there is no boar present she sniffs and nuzzles at the genitals of other sows, but if there is a boar she promptly accepts mating and in the normal course of events becomes pregnant. The cycles then cease until the young are born. If she does not mate, or if the mating is not fertile, cycles continue as usual every 3 weeks throughout the year.

These events in the life of the sow have been known to every swineherd for ten thousand years, but nobody knew until the present century just what is going on inside the animal every 21 days to stir her two hundred pounds of meat, bones, and fat into three days of such specialized conduct. It turns out, when we investigate the matter, that during the diestrous phase of the cycle (the $2\frac{1}{2}$ weeks of no sexual activity) the Graafian follicles in the ovaries are all small, with unripe eggs in them. About two days before the onset of estrus, a crop of follicles begins to grow. The ovaries of sows killed on the first day of estrus contain large follicles with mature eggs. Late on the second day we find that the follicles have ruptured and the eggs are on their way down the oviducts. The follicles are being converted into corpora lutea (see diagram, Fig. 15). By the sixth or seventh day after ovulation the corpora lutea are fully developed and (as we shall see in Chapter V) are at work producing their hormone, progesterone. This hormone acts upon the uterus, altering its lining to make it ready to receive the eggs, as a plowman tills the fields to receive the seed. If the sow has mated while she was in estrus, the eggs will be fertilized and they will settle down in the uterus and develop there. Once the pregnancy is well established, the cycles cease, possibly because a hormone produced by the placenta (or rather the outer part of the embryonic tissues, destined to form the placenta) signals the ovary, via the pituitary gland, to stop development of follicles for the time being. If not fertilized, the eggs will degenerate and go to pieces; then the corpora lutea, no longer useful, begin to

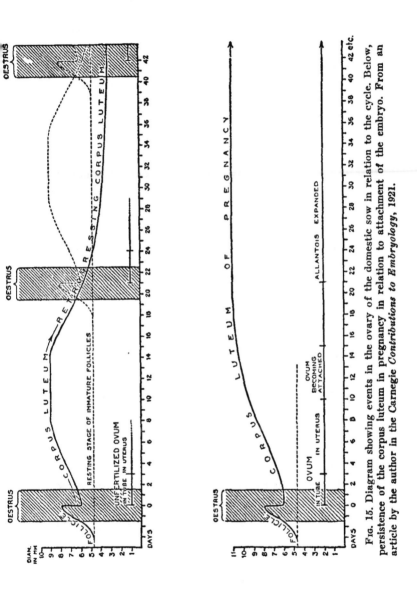

FIG. 15. Diagram showing events in the ovary of the domestic sow in relation to the cycle. Below, persistence of the corpus luteum in pregnancy in relation to attachment of the embryo. From an article by the author in the Carnegie *Contributions to Embryology*, 1921.

{ 67 }

degenerate on the 15th day of the cycle (i.e. 14 days after ovulation) and shrink out of existence. About the 19th day a new crop of follicles begins to mature and the cycle repeats itself.

The whole process of the estrous cycle is therefore a beautifully timed arrangement by which, first, the eggs are matured and discharged from the ovary; second, the sow is induced to mate at just the right time to get the eggs fertilized; and third, the uterus is prepared to receive the embryo, by action of the corpus luteum, which is formed and thrown into action at just the right time. If all this preparation for maternity fails for lack of opportunity to mate, then the cycle repeats itself as soon as the changes in the ovary and uterus have cleared away.

Variations of the cycle in other mammals. What I have described in the sow is the fundamental plan of the cycle; this animal happens to illustrate it with diagrammatic simplicity. A whole book could be filled with variations displayed by various animals.[3] In guinea pigs, for example, the female not only will not mate between estrous periods, but actually cannot, because the skin grows over the vaginal orifice and blocks entry of the male, except during estrus, when it temporarily opens. Such an arrangement exists in no other animal. In cats and ferrets the ovarian follicles behave peculiarly; although they ripen in each cycle, they will not rupture and discharge their eggs unless mating occurs. In rabbits the follicles will not even ripen without mating. This means that in these three species the corpus luteum phase of the cycle does not occur if the animal does not mate. In all other animals known at present, including man, the follicles ripen and rupture spontaneously.

In rats and mice there is a very peculiar situation, first

[3] See S. A. Asdell, *Patterns of Mammalian Reproduction*, Ithaca, N.Y., 1946.

worked out by Long and Evans.[4] The cycles are rapid, averaging less than 5 days in length. Since it takes about 7 days to get the embryos down into the uterus and safely implanted there, we see that unless something were done to prevent it, there would be two batches of early embryos on their way at once. Before the first were soundly attached, the second lot would be claiming space and nourishment in the uterus, with resultant confusion and damage. To prevent this a special mechanism has developed in rats and mice: the act of mating signals the ovary to postpone the next cycle 10 days instead of five, thus giving time to get the pregnancy under way. This can be imitated experimentally by simply inserting a smooth glass rod deeply into the vagina during estrus, in lieu of the male organ.

The human cycle: menstruation. The most peculiar variation of all occurs in the human species and in our near kin, namely the apes and higher monkeys. The length of the cycle is about 4 weeks, but in these animals there is no well-defined estrous period. Mating can occur at all times of the cycle.

The follicle matures and discharges its egg silently, without any marked changes of behavior. The corpus luteum forms from the discharged follicle and functions as in other animals, but when it breaks down, about two weeks after ovulation, its effect upon the lining of the uterus does not merely subside. There is, instead, a sharp breakdown of the endometrium with hemorrhage. This periodic *menstruation* occurs only in the higher primates; nothing like it is seen in other animals. The question of its relationship to the estrous cycle of non-menstruating animals has puzzled and confused naturalists and physicians since the days of Aristotle. Because animals like the sow and mare have in their cycles one prominent event, namely estrus, and humans display also one definite cyclic

[4] J. A. Long and H. M. Evans, "The oestrous cycle in the rat and its associated phenomena." *Memoirs of the University of California,* vol. 6, 1922.

change, namely menstruation, the two phenomena were thought to be the same. Serious misconception of the human cycle caused by this error has been cleared up only in the twentieth century.

For the sake of perfect clearness on this point, let us compare diagrams of the estrous and the menstrual cycles (Fig.

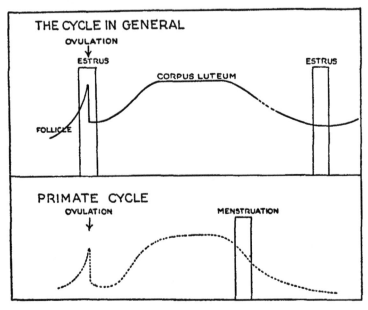

Fig. 16. Diagram comparing the estrous cycle in general with the menstrual cycle of the higher primates.

16). It will be seen that the two are fundamentally alike, since the important feature of each is ovulation followed by the corpus luteum phase. In lower animals ovulation is accompanied by an outspoken period of sexual receptiveness, in the primates (man, apes, and higher monkeys) it is not; in the primates the end of the corpus luteum phase is accompanied by menstruation, in the other animals there is no bleeding at this time.

How menstruation is brought about by the ovarian hormones, and what purpose it may serve, we shall discuss in Chapter VI.

The vaginal cycle. Obviously, the periodic ripening of the ovarian follicles sets in action large forces that can alter the state of other organs, change the reactions of the body, and determine the behavior of the female animal. By means of the ovarian hormones released at this time, there are (as we shall see) cyclic changes in the whole reproductive tract. Not only is the lining of the uterus modified, but the Fallopian tube, the vagina, and in some species even the external genital organs also take part in these rhythmic alterations. This fact led to the discovery, twenty-five years ago, of an extraordinarily useful method of research, which greatly increased our rate of progress in unraveling the problems in this field; and because it was made in the United States, helped put the American investigators of this generation in the forefront.

To grasp the importance of this discovery, we must remember in the first place how very helpful to science in general are the rat, mouse, and guinea pig. These little rodents are hardy, inexpensive, and easy to feed, house, and handle in the laboratory. Their small size is also a great advantage when experiments call for treatment with scarce or expensive hormones and other chemical reagents. Unlike cats, they breed freely in cages. Unlike dogs, they have rapid cycles, requiring no long waits in the course of experimental study. Unfortunately, among all the mammals they give the least conspicuous signs of estrus. They show no genital swellings like the bitch, no excitement like the sow. They normally mate only at night, and to be certain they are in estrus, the investigator had to put them with males and sit up all night watching them, or else use roundabout methods of observation which were not perfectly reliable. As an illustration of the difficulties, I recall that before 1917 a distinguished biologist who was working chiefly with guinea pigs supposed their cycle to be 21 days instead of 15, and a first-class embryologist who made a determined effort to work out the estrous cycle of the rat by the best means at his command, came out with a result of 11 days, just twice as long as the correct figure, having somehow missed alternate cycles.

It can be imagined with what enthusiasm those of us who were working in the field learned of the simple discovery announced in 1917 by C. R. Stockard of Cornell Medical College and his colleague G. N. Papanicolaou.[5] These men found that in the guinea pig the cyclic changes, which take place in the reproductive tract under the influence of the ovarian hormones, are seen with especial clearness in the lining of the vagina. Here there is a cycle of growth and shedding of the surface cells, which follows events in the ovary so closely that the time of rupture of the follicle can be determined within one hour. It is only necessary to scrape the vaginal lining gently with an instrument, or to wash the cavity out with a medicine dropper and a little salt solution, and study the findings under the microscope. This can be done in a few minutes and does not harm or upset the animal in any way. The vaginal closure membrane, mentioned on page 68, as a peculiarity of the guinea pig, is of course kept open by these daily examinations.

In justice it ought to be added here that something of what Stockard and Papanicolaou found had been described in the 1890's in more or less forgotten papers by several investigators, particularly the French observer, Lataste; but it was their masterly and complete investigation of the matter which made it available to science.

By the use of this method Stockard and Papanicolaou promptly ascertained the correct length of the guinea pig's cycle and secured a much more accurate timing and description of the subsequent events of the cycle than we had before.

[5] C. R. Stockard and G. N. Papanicolaou, "Oestrous cycle in the guinea pig," *American Journal of Anatomy*, vol. 22, 1917.

The method was soon applied to the rat by Long and Evans of the University of California and to the white mouse by Edgar Allen of Yale Medical School, then at Washington University, St. Louis. We shall see in subsequent chapters some of the important work that was made possible because the cycles of these three animals were now so much better known. The detection of the ovarian estrogenic hormone by E. Allen and Doisy (1923) was a direct result; so was the discovery of Vitamin E by H. M. Evans and his colleagues (1922). In conjunction with the growing knowledge of the cycle of the sow and of reproduction in the rabbit, it led indirectly to the discovery of the corpus luteum hormone and to a far clearer general comprehension of the whole field than was possible before.

The exact details of the vaginal changes are of no particular importance except to those who need to follow them in the laboratory. Briefly stated, there are two kinds of cells that get free in the vagina. One kind is of course the cells of its lining. These are epithelial cells like those of the external skin except that they are ordinarily moist. Those that lie on the surface of the lining are not infrequently shed into the cavity. The cyclic changes can be followed in the accompanying figure, taken from the monograph on the rat's cycle by Long and Evans. In Plate XIII, *B* we see the vaginal cells as they are in the interval or diestrous phase; there are fairly large numbers of white blood cells or leucocytes (the small rounded cells with irregular nuclei) intermingled with a few large flat epithelial cells. At *C* we see signs of approaching estrus; the leucocytes have disappeared, and the epithelial cells are swollen and rounded, by action of the ovarian hormone upon them before they were shed from the wall into the cavity of the vagina. In figure *D* of Plate XIII we see that the epithelial cells now being shed are dry, scaly, and lack nuclei —they are cornified, as we say, like the cells on the dried-out surface of ordinary skin. At the time when the ripe follicles are about to rupture, the process of cornification of the vaginal wall is very active, and cornified cells are shed in thousands, yielding thick caseous scrapings that can be identified without a microscope. The next day the leucocytes come back (E), the masses of cornified cells disintegrate, and the interval picture reappears. In sexually mature rats and mice this change repeats itself every 5 days, in guinea pigs every 15 days.

Vaginal cycles in other animals. If such clear-cut vaginal changes occurred in all mammals, it would be of great advantage in studying their cycles. If they occurred in women, it would be of priceless value to gynecologists, especially in the serious business of diagnosing the cause of sterility. At present we have no certain way of finding out whether a woman is ovulating or not, except by operative exploration. Unfortunately, the vaginal changes are far less clear than in the small rodents. Extreme differences, such as that between normal activity of the ovary on one hand and total inactivity on the other, can be detected, but such changes as occur from day to day in the cycle are faint. The latest word thus far is in volume 31 of our Carnegie *Contributions to Embryology*, recounting studies made on human patients by Ephraim Shorr of Cornell Medical College and observations of the Rhesus monkey made in the Department of Embryology of the Carnegie Institution by Inés de Allende in consultation with Carl G. Hartman. Dr. de Allende was generally able to diagnose the occurrence or nonoccurrence of ovulation in monkeys, as confirmed by surgical exploration. It seems not improbable that by refining the study of vaginal cells it may ultimately be possible to diagnose ovulation in humans.

The cause of the cycle. Some day no doubt we shall understand the whole mechanism of the ovarian rhythm, and know why and how the cycle is short or long in various animals and why its manifestations vary in so many ways. At present we can do hardly more than guess. We think the alternations of

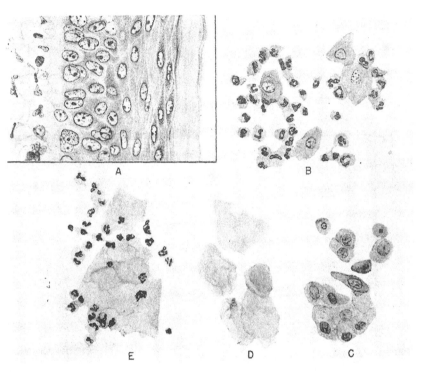

PLATE XIII. The vaginal cycle in the white rat. *A*, section of lining of the vagina. The inner surface, from which cells are shed as seen in following pictures, is at the right. *B*, cells shed into cavity of vagina during interval part of cycle. The larger masses are the epithelial cells; the small rounded objects with gnarled nuclei are white blood cells (leucocytes). *C*, just before estrus. *D*, during estrus (*at time of ovulation*) *showing complete cornification*, with loss of nuclei. The section of the wall of the vagina pictured at *A* was taken at this time, and shows the cornified cells in place, forming a clear layer at the right. *E*, at end of estrus, showing return of leucocytes. All magnified about 650 times. From J. A. Long and H. M. Evans, "The Estrous Cycle in the Rat," *University of California Memoirs*, No. 6.

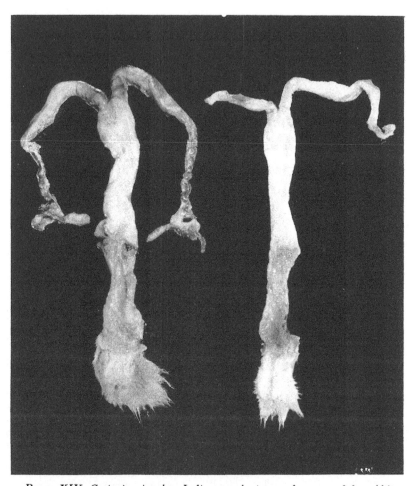

PLATE XIV. Castrate atrophy. *Left,* normal uterus of young adult rabbit. *Right,* uterus of litter-mate sister one month after removal of both ovaries. Note thinner horns and vagina, with paler and flabbier walls, in specimen on right. (One Fallopian tube of this specimen has been cut off.) Preparation by author. Natural size.

the cycle are due to an interplay or see-saw action between the hormones of the ovary and of the hypophysis or pituitary gland. To make this clear let us continue to anticipate the next chapter by postulating that the ovary produces an estrogenic hormone that acts upon the uterus and the rest of the reproductive tract, producing among other effects the estrous changes of the vagina referred to above. Another fundamental fact is that the whole activity of the ovary, including the production of the estrogenic hormone, is under the control of the pituitary gland (see Plate XIX and Fig. 20). If this remarkable bit of glandular tissue is removed, the ovary quits functioning altogether. By chemical extraction the pituitary yields hormonal substances which can restore ovarian function in the absence of the pituitary gland, or cause the ovary of the immature animal to grow and begin functioning. It has been shown, however, that if we inject the ovarian estrogenic hormone into a female animal, the pituitary hormones that stimulate the ovary are reduced in amount. Here we have, in all probability, the fundamental mechanism of the ovarian timepiece. The pituitary stimulates the ovary, but the latter then sends out its estrogenic hormone and this depresses the pituitary. Then the ovary loses pituitary support. Up goes the pituitary again like the other end of a see-saw. There must of course be other factors in the situation, for such a mechanism, like the see-saw, will come to balance unless it gets a push from time to time. In Chapter VI we shall return to this subject after a more detailed account of the ovarian hormones.

THE HORMONE OF PREPARATION AND MATURITY

"*The hierarchy* [*of the organs*] *is such that incessantly one borrows from another, one lends to the other, one is the other's debtor . . . each member prepares itself and strives anew to purify and refine this treasure . . . each doth cut off and pare a portion of the most precious of its nourishment; and dispatch it downwards where Nature hath prepared Vessels and Receptacles suitable . . . to preserve and perpetuate the Human Race. All this is done by Loans and Debts from one to the other.*"—RABELAIS, *Pantagruel* (Book III, Chapter 4).

CHAPTER IV

THE HORMONE OF PREPARATION AND MATURITY

THE ovary was very slow to yield the two great secrets of its function. The fact that mammals and man breed by means of eggs, and that the ovary is the source of the eggs, was not even conjectured until 1672, and was not proved (as we have seen, p. 34) until 1827. The fact that the ovary is an organ of internal secretion was not clearly stated until 1900.

The ancients knew of course that the organs we now call the ovaries are homologous with the male testicles and have something to do with reproduction; in fact, the Greeks and Romans called them "the female testes." Castration of female animals to prevent them from breeding is a very old practice. The verb "to spay," meaning to castrate the female, goes back to late Middle English, and practitioners of that art were called "sow-gelders" as early as 1515, judging from a citation in the *New English Dictionary*. These men must have known that removal of the ovaries stops the estrous cycles, and anybody who butchered a spayed sow would surely notice that in the absence of the ovaries the uterus shrinks far below its normal size. But these facts, even if known from observations on animals, did not get into the textbooks of human physiology until the surgeons began to remove human ovaries. That operation, first made possible in 1809 by the courage of Ephraim McDowell and of his patient, Jane Crawford, was fairly common by 1850. The great physiologist Carl Ludwig said, in his textbook of 1856, that in humans loss of the ovaries not only stops the menstrual cycles, but also causes the uterus to shrink.

This matter of *castrate atrophy* furnished a really important clue. It deserves careful explanation. When the ovaries of an adult female are removed, the oviducts, uterus, and vagina undergo rapid reduction in size. In the rabbit, in which the phenomenon has been quite fully studied, the uterus loses half its weight in two or three weeks and its tissues become thin and flabby (Plate XIV). This atrophy of the uterus produced by castration of the adult female is obviously just the reverse of what happens at the time of puberty, when the ovaries first become mature. The uterus and the other accessory reproductive organs (oviducts, vagina and mammary glands) which in infancy are small and undeveloped, grow to adult size when the ovaries begin to function. Both pubertal growth and castrate atrophy indicate clearly that the adult uterus is dependent upon the ovary.

It is difficult to realize that less than fifty years ago nobody could guess how this action of the ovary upon the uterus is produced. Some thought that castrate atrophy was due to interference with the blood supply of the uterus when the ovaries were removed, others thought the nerve connections were upset.

A Hormone of the Ovary?

The first step toward demonstrating the endocrine function of the ovary was taken in 1896 by Emil Knauer of Vienna, who took out the ovaries of guinea pigs and grafted pieces of them back into the same animals, at new sites. He demonstrated that such grafted ovaries prevent the occurrence of castrate atrophy. It must be, then, that the ovary makes some sort of chemical substance which acts upon the uterus. This is lost when the ovaries are removed, but again becomes available if the ovary is successfully regrafted, no matter where, in the animal's body.

In 1900 Knauer definitely suggested this idea of an internal secretion of the ovary. In the same year Josef Halban, also of Vienna, took three infantile guinea pigs, grafted bits of adult ovaries of the same species under their skin, and found that their uteri promptly grew to adult size. On the basis of

Knauer's work and his own he stated the hypothesis of an internal secretion in perfectly clear terms: "We must assume that a substance is produced by the ovary, which when taken into the blood is able to exert a specific influence upon the genital organs; and that the presence of this substance in the body is absolutely necessary for the maintenance—and, as my researches show, for the development—of the other genital organs and the mammary glands."

With such a downright challenge as this, the inevitable next step was for somebody to try to make a chemical extract of ovarian tissue which should contain the potent substance postulated by Halban, and which could be injected into castrated animals instead of grafting ovaries into them. Several investigators actually tried it, but they were working in the dark and failed to hit upon the proper chemical steps. Thereafter for a few years experimenters went off on another trail. From 1911 to 1914 several Viennese and German gynecologists spent a great deal of time and effort searching for chemical extracts of the ovary which should produce menstruation in animals. This was a bad idea, for apparently these doctors never stopped to think that the rabbits and guinea pigs they were using do not menstruate anyway. We can see now that if they had used monkeys, which do menstruate, they might have found a clue. As a matter of fact, these men—Adler, Aschner, Schickele—did get a clue, but not exactly what they were looking for. They did not accomplish the miracle of making guinea pigs menstruate, but many of their extracts did have the property of increasing the blood flow through the vessels of the immature uterus, thus making it grow. Unfortunately they did not all use similar methods, and, what was more confusing, some of the extracts were made from whole ovaries, some from the corpora lutea (of swine or cows) and some even from human placentas. The situation was so confused that the results were almost meaningless.

In 1912 and 1913 two workers, Henri Iscovesco in Paris

and Otfried Fellner in Vienna, found that a really potent preparation could be made by extracting the ovary with fat solvents (alcohol, ether, acetone). Their products readily prevented castrate atrophy, developed the infantile uterus, and enlarged the mammary glands. In the last year or two before the War of 1914-1918 these results were refined and standardized by several workers, the best work being that of Robert Frank of New York and Edmond Herrmann of Vienna.

The sum total, then, of twenty years of investigation was the demonstration that there is a substance in the ovary in general, in the corpus luteum, and in the placenta, which has the property of causing growth of the uterus of the infantile animal and of preventing castrate atrophy in adult animals. The relationship of this substance to the estrous cycle and to menstruation was decidedly a problem for the future, and its presence in so many tissues hindered rather than helped the effort to untangle the specific endocrine functions of the ovary and of the corpus luteum.

The Vaginal Test; Isolation of the Hormone

During the war the European laboratories dropped such problems as this, but in the United States the discovery (or rather, rediscovery and clarification) of the vaginal cycle of rodents by Stockard and Papanicolaou turned the work in another direction. I mentioned in Chapter III that Edgar Allen in 1922 described a similar cycle of vaginal changes in the mouse. Allen was very much impressed by the striking coincidence of the peak of the vaginal changes with the presence of mature follicles in the ovary. This led him to consider the possibility that there is a hormone in the follicles and particularly in the follicle fluid. The hypothesis that the special events of estrus are due to a secretion of the follicle, rather than of some other element of the ovary, was already widely, if somewhat vaguely entertained, because various observers had noticed that the mature phase of the follicles is closely associated with the phenomena of estrus. The pioneer American worker, Leo Loeb, said in 1917 that some of the cyclic changes in the uterus might be due to a secretion of the follicles. Arthur Robinson of Edinburgh in 1918 went so far as to write "It can scarcely be doubted that the phenomena of heat are due to something produced by the follicles." Edgar Allen now had the added evidence of the vaginal cycle pointing him in the same direction. Enlisting the collaboration of Edward A. Doisy, then a young biochemical colleague at Washington University (St. Louis), he proceeded to test this hypothesis by injecting a few drops of fluid, drawn from mature follicles of the sow, under the skin of castrated female rats and mice. In such animals, of course, cycles do not occur, and the vaginal lining becomes very thin and undeveloped and remains unchanged from day to day. An injection of follicle fluid, however, produced in 48 hours a typical estrous condition of the vagina, which could be easily detected by scraping or washing out the vagina and looking at the cells under the microscope. This result is shown in Plate XV, originally published in illustration of Allen and Doisy's earliest results. When administered to infantile rats and mice, the injections caused the uterus to grow to adult size and the vagina to open as in sexually mature animals (see Plate XVI, illustrating similar growth of the uterus in monkey and rabbit).

This astonishing result provided at once a new test, simple, precise, and rapid, by which the chemists could follow the hormone through various steps as they attempted to purify it. A sample suspected to contain it can be injected into a castrated rat or mouse and the vaginal cells examined under the microscope at intervals until the estrous change is seen. By giving graded doses to a series of animals the amount of hormone in a sample can be measured.

Follicle fluid is a complicated mixture of water, salts, a little fat, a lot of protein. Doisy faced the task of extracting from this "soup" a substance that could be present, as he knew well, only in very slight amounts. He guessed that like Iscovesco's, Fellner's, and Herrmann's materials it might be soluble in fat solvents. Fortunately this was correct, and he found at once that the potent substance is soluble in alcohol, a very helpful thing because when strong alcohol is added to a beaker of follicle fluid it throws down the troublesome proteins but leaves all the potency in the clear supernatant fluid, which can be decanted and subjected to further analysis. The hormone is also soluble in ether, chloroform, and acetone. In this respect it resembles the fats and therefore goes along with them through the various solutions; but Doisy was able to get rid of these substances by the familiar expedient of cooking the extract with a little alkali. The fats were thus converted to soaps and could be washed away with water, leaving a small amount of non-saponifiable oil, in which was now concentrated almost all the potency of the original follicle fluid.

For theoretical reasons not now important, Allen and Doisy believed their active principle was peculiar to the large ovarian follicles, but other workers soon found it in the rest of the ovary. There is a little in the corpus luteum, and very much in the placenta. Robert Frank and his colleagues found it in the blood of female animals. It even turned up in the sex organs of plants, for example in the catkins of willow trees and in palm kernels.

Incidentally, it was now obvious that this *estrogenic hormone* (as we may call it, because it initiates the estrous

PLATE XV. Effects of the estrogenic hormone on the vagina of the rat. *A* section of vagina 10 days after castration. *B*, cells from contents of vagina at this time; leucocytes only are seen. *C*, section of vagina of castrated animal 36 hours after an injection of estrogen. Note thickening of vaginal lining *D*, cells from vaginal contents of same animal. Epithelial cells now predominate in the vagina. *E*, section of vagina in estrous condition 48 hours after beginning of treatment with estrogenic hormone. Surface cells cornified, forming clear surface layer. *F*, vaginal cells of same animal, showing cornified cells only Greatly magnified. From *Sex and Internal Secretions*, by courtesy of Edgar Allen and the Williams and Wilkins Company.

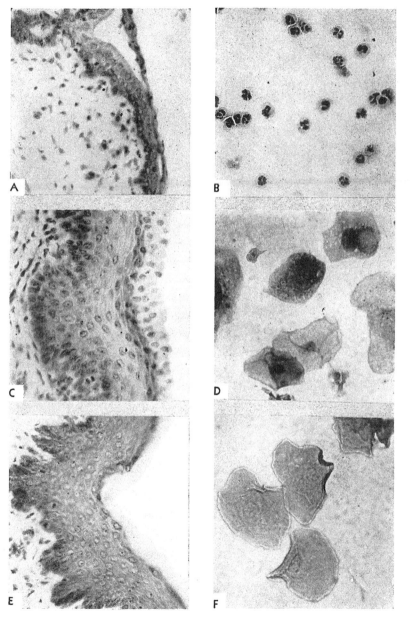

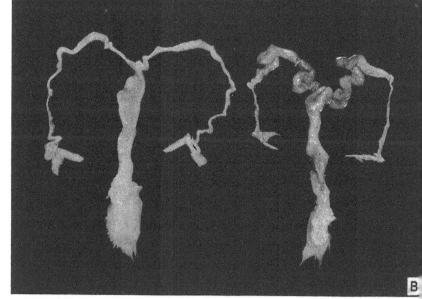

changes in the vagina) is the very same substance that Halban had predicted and that Iscovesco, Fellner, Frank, Herrmann and many others had extracted, in varying degrees of impurity, from ovaries and placenta. It not only acts upon the vagina and makes the immature uterus grow with remarkable speed, but it also stimulates growth of the external genital organs and of the mammary glands.

Complete chemical purification proved very difficult. After collecting a supply of follicle fluid by tapping hundreds of ovaries, or grinding up a batch of ovaries or placentas, extracting with alcohol and putting the extract through twenty chemical steps, it was disheartening to find that various oily contaminants, some known and some unknown, insisted upon following the hormone through all the processes of purification.

At this point another unexpected discovery came to the rescue. In two European laboratories (Loewe, 1926, and Aschheim and Zondek, 1927) it was found that a substance having the same potencies is excreted in the urine of adult human females. Especially during pregnancy, when the hormone is present in the placenta in large quantities, it passes through the kidneys and into the urine in large amounts. Investigators now began to look for it in the urine of other animals, and Aschheim made the almost incredible discovery that it is present in enormous amounts in the urine of stallions, surely the least feminine of animals. The reasons for this strange fact

PLATE XVI. *A*, cross sections of uteri of immature Rhesus monkeys showing e effect of estrogenic hormone. *Left*, untreated animal. *Right*, treated animal. ote growth and differentiation of tissues. Magnified 8 times. From the *Journal Morphology*, by courtesy of Edgar Allen and the Wistar Institute of natomy and Biology.

B, uteri of immature rabbits, showing effect of estrogenic hormone. *Left*, treated rabbit, 6 weeks old. *Right*, litter-mate sister after 10 days' treatment. ote growth (thicker horns), increased circulation of blood (shown by darker lor of horns), and improved muscle tone (shown by coiling of the horns). 8 natural size. Preparation by author.

may be put aside for the moment; the important thing is that a watery source of the hormone is very much easier to work with than follicle fluid or chopped-up placenta. Starting with human pregnancy urine or stallion's urine, the biochemists did not have to contend at all with fats and proteins, the most troublesome ingredients of their former sources of supply. When the kidney secretes urine it strains out and retains these substances in the body. With the aid of this great simplification, Doisy was able to announce in 1929 that he had obtained the active principle in crystalline form, that is to say, absolutely pure; in the same year the great German biochemist Butenandt produced it and in 1930 it was announced at Amsterdam by Dingemanse, de Jongh, Kober and Laqueur and from Denver by d'Amour and Gustavson. All four laboratories had found exactly the same substance.

The Chemical Nature of Estrone

It was now up to the organic chemists to tell us the chemical nature of this hormone from pregnancy urine, or estrone,[1] as it came to be called. What they found, I shall have difficulty in stating for my readers, except those who are familiar with the chemistry of hydrocarbons, because estrone belongs to a group of substances not within the ken of the general public, nor even indeed, of many chemists. It is a *sterol*. To explain this by saying that sterols are complex higher cyclic alcohols is correct but not very helpful. They are colorless solids occurring in crystals; in bulk they look a good deal like powdered sugar or table salt, but when compressed into a

[1] The term *estrogen*, or *estrogenic hormone*, has been adopted to signify any and all of the hormones giving effects such as described in this chapter. *Estrin*, now little used, has the same significance. The terms *estrone, estradiol, estriol, stilbestrol, equilenin*, etc. refer to individual hormones of the series, each of which has its own particular chemical formula. Allen and Doisy, the discoverers of estrone, called it *theelin*, and this name, with its derivatives, is by agreement of a committee of American scientists alternatively used in this country. Each drug manufacturer making these hormones has his own trade name.

mass, a quantity of sterol looks and feels more like a hard white crystalline wax. Sterols are plentiful in yolk of eggs, brain tissue and many plant tissues. Wool fat (lanolin) is largely composed of sterols mixed with softer, greasier fatty substances. Most of the sterols have no hormone action; in fact they tend to be rather inert substances, but since the identification of estrone it turns out that several particular sterols that are found in the male and female reproductive systems and the adrenal gland, as well as related substances prepared synthetically, have powerful effects in the animal body. Estrone is one of this group.

Readers who are interested in the chemical structure of estrone and the other sex gland hormones will find in the Appendix of this book a detailed account, written for those who remember a little of their elementary chemistry. What follows here will be intelligible to those who are familiar with organic chemistry; other readers are advised to read Appendix I before proceeding further. The structural formula of estrone is as follows:

ESTRONE

The molecule contains 18 atoms of carbon, 22 of hydrogen, 2 of oxygen, arranged as 3-hydroxy, 17-keto estratriene. In 1936 Marker, Kamm, Oakwood and Laucius gave us absolute and final proof of this structure, by producing estrone artificially in the laboratory (at Pennsylvania State College) by conversion of one of the sterols obtained from vegetables, of which the formula was already known. This was of course a partial synthesis or rearrangement, the investigators having taken advantage of Nature's work in building up the substance with which they started. Bachmann, Cole and Wilds of the University of Michigan reported in 1939-1940 that they had achieved the total synthesis of an estrogenic hormone,[2] building it up in the laboratory from simple materials. These examples of chemical magic made a fitting climax to years of brilliant work on estrogenic hormones by the chemists of many nations.

A substance as complicated as this affords many opportunities for slight modifications by rearrangement, addition, or subtraction of the constituent atoms. It is not surprising, therefore, that a whole series of estrogenic hormones has been found, each differing slightly from estrone in chemical structure and in potency or even in details of physiological action. These have been obtained from the urine of other species (as for example *equilenin* from the mare), from male urine and from the human placenta.

It is a curious fact that Allen and Doisy's originally discovered hormone of the swine ovary, being immensely difficult to purify because of all the fats, oils and proteins of the tissues that come out in the extract with the hormone, was not actually identified for a long time. Finally MacCorquodale, Thayer and Doisy in 1935 extracted two tons of ovaries and obtained as the chief active substance a few milligrams of a compound differing from estrone only in having an OH group at position 17 instead of the doubly-bonded oxygen. This is *estradiol*. It is probably the form which is actually made in the ovary.

It has been found that esters of these hormones, i.e. combinations of estrogens with organic acids, are more slowly eliminated from the body than the estrogens themselves and therefore give longer and more intense effects. The propionic and benzoic esters have been widely used in the treatment of human patients.

Chemists are always interested to know just what part or feature of such a molecule is responsible for its effects. They

[2] Actually not estrone, but equilenin, referred to in the following paragraph.

learn this (if they can) by making up similar but definitely different substances until they find the simplest substance that has the same action. E. C. Dodds of the Courtauld Institute of Biochemistry, Middlesex Hospital, London, and his fellow workers have thus found a long series of synthetic estrogens, of which the most important is *diethyl stilbestrol*:

DIETHYLSTILBESTROL

This compound is practically identical in its effects with the naturally occurring estrogens and is being tried by physicians in place of them. Students of chemistry will be interested in the fact brought out by Dodds, that the formula of diethyl stilbestrol can be written so that it resembles, in spatial relations, the formula of estrone and the other natural estrogens.

DIETHYLSTILBESTROL

The outstanding difference is that in stilbestrol two of the rings are open. Dodds compares this sort of similarity to that of a particular key, made for a given lock, with the skeleton key which will also open it. When, however, the molecules of diethyl stilbestrol and estrone are represented by 3-dimensional models, as was kindly done for me in connection with the Vanuxem Lectures, by Mr. Muller of the Princeton Chemistry Department under the direction of Professor Hugh Taylor, they do not show the same degree of similarity as do the two diagrams on paper. The relation between the chemical structure and the action of these substances can hardly, therefore, be explained by a hypothesis as simple as that suggested by Dodds. All we can say is that all the estrogens thus far known, both natural and synthetic, have a ring structure more or less like these examples and possess phenolic properties, as indicated by the OH group in at least one ring. Until we know definitely just what the hormones do when they reach the cells upon which they act, these facts will have to suffice us.

Potency of estrone. Estrone is a very powerful substance; that is to say, only a very small amount is required to produce large effects. It has to be used and talked about in quantities too small for ordinary measures of weight, and so we mention it in terms of the chemists' tiny unit, the gamma (γ) which is 1/1,000 of one milligram. An ordinary U.S. postage stamp, gummed, weighs 60 milligrams; one gamma is therefore 1/60,000 part of the weight of a stamp. As little estrone as 1/100 gamma, or 1/6,000,000 of a postage stamp, per day for 3 or 4 days may be enough to produce estrous changes in a castrated mouse, and even in a woman 20 gammas per day (1/3,000 of a postage stamp) for 10 days may be sufficient to produce certain profound effects upon the uterus which we are going to consider in Chapter VI. Some of the other estrogens are three or four times as potent as this.

The total amount produced in one day by the two ovaries of an adult woman is believed to be equal in potency to about 300 to 400 gammas of estrone, or something over 1/200 of the weight of a postage stamp.

By worldwide agreement through the League of Nations an international unit of estrone was set up, for the benefit of scientists and the drug industry. This is the amount of potency in 0.1 gamma of the pure hormone. At first the League of Nations Committee on Standardization of Drugs and Hormones set aside in London a stock of the pure hormone to serve in much the same way as the international standards of weight and length (metric system) in Paris. Any responsible person who wished to test his own product could get a few units from London and compare the effectiveness of the two samples in rats or mice. It is good to know that if the precious store of standard estrone should ever be lost the hormone can be recreated to the exact formula and as long as the powers of destruction leave us a good organic chemist alive on earth this product and symbol of constructive international effort will survive.

Administration of the hormone. Most of the pure estrogens must be administered hypodermically, because they are not well absorbed from the digestive system and are therefore not effective when given by mouth. Some of them can be dissolved in water for hypodermic injection, but the effects last longer when they are injected in oil. For this purpose bland vegetable oils such as corn (Mazola) oil and sesame oil are used. The neutral triglycerides are also coming into use as solvents for injection; they dissolve the hormones better than do the oils and are generally better tolerated by the tissues into which they are injected. It has been a great surprise to find that the oil-soluble sterol hormones will enter the body through the skin if made up into ointment and applied by thorough rubbing (inunction). This is not a very exact way of giving a drug, for we can never be sure how much is lost or not absorbed, but at least it avoids disagreeable hypodermic punctures. Some of the new synthetic estrogens of relatively simple form, such as stilbestrol (mentioned above) can be given by mouth. There are also certain new compounds of the higher estrogens (e.g. ethinyl estradiol) which are effective by mouth. We may expect that after a due period of experiment, injections of the hormones will give way, at least to a considerable degree, to medication by mouth and by inunction.

The English investigators Deanesly and Parkes suggested, a few years ago, a brand new method of administering the estrogenic hormones (and equally well the other sterol hormones, whether from corpus luteum, testis, or adrenal gland, or from the chemist's beaker). This is to compress the hormone into a little rod-shaped pellet like a short segment of graphite from a lead pencil, and to bury this pellet under the skin. A hypodermic needle of large diameter can be inserted and the pellet pushed through its canal. Once buried, the surface of the pellet is slowly dissolved and the hormone is thus received by the animal in very small but continuous dosage. In this way it acts very effectively over long periods of time with the minimum disturbance of the animal. The method is extremely useful in laboratory experiments and there are reports of its successful use in certain human cases. There is still much to be learned, however, about the rate of absorption of these pellets, before we can know what daily dose we are actually giving when we administer the hormone in this way.

What part of the ovary makes the estrogenic hormone? After a great deal of investigation it is generally considered by the experts that the estrogenic hormone is probably made in the walls of the follicles, both large and small. The evidence for this conclusion is indirect, for there is no way to locate the hormone directly. The best we can do is to put little fragments, taken from various parts of the ovaries of large animals, under the skin of castrated mice to see if they contain estrogenic potency. Some years ago A. S. Parkes of London astonished his fellow workers by reporting that he had found a way (by use of X-rays) to break down the follicles in the ovaries of mice and thus to reduce the ovaries to masses of nondescript cells. Such mice became sterile, of course, because they could not produce eggs, but they went on having estrous cycles more or less regularly. This teaches us that the ovarian cells which make the estrogenic hormone can do so even when not actually organized into follicles.

I have already mentioned that during pregnancy the human placenta contains a large amount of estrogenic hormones. We know that these are not made in the ovary but in the placenta itself. There are several cases on record in which the ovaries were removed during pregnancy but the supply of estrogen continued. In humans and animals that are not pregnant the ovaries are the only effective source of estrogen, and if they are removed all evidence of estrogenic action disappears.

Recent investigations begin to hint that under special circumstances the adrenal glands may produce estrogenic hormones. It is too early however to see what relation this has to the general theory of the estrogenic hormones.

Action of the Estrogenic Hormones

Once the hormone gets into the blood stream, whether from the subject's own ovaries, or by the experimenter's needle, it is carried all over the body, through all the organs and tissues. But—and this is characteristic of all the hormones—it is selective in its action, like a key that opens certain locks and not others. The major point of attack of this particular hormone is the reproductive tract. This is strikingly seen if we give small doses to an immature female animal, whose uterus has never yet been subject to the action of estrogen. Half an hour after an injection of the hormone the blood vessels in the uterus begin to dilate and the blood to flow faster through them, reddening the whole organ. The microscope tells us that the rate of cell multiplication in uterus, Fallopian tube and vagina is sharply increased. All the constituent tissues of these organs become better defined or differentiated, as the histologists say (see Plate XVI, *A*). The muscle cells of the uterus grow longer and thicker and the uterine glands increase in size. Plate XVI, *B* shows the remarkable stimulation of growth and intensification of blood supply, produced in a young rabbit, 10 weeks old, by treatment with estrogen for 10 days. In two weeks or thereabouts, if the treatment is continued, the baby rabbit develops a uterus of adult size; but to the best of my knowledge, the hormone will not carry the growth of the uterus beyond the normal mature state. Further treatment, if forced, might cause damage to the uterus but would not make a larger-than-normal organ. The old hereditary pattern of the body does not yield readily to these upstart hormones.

As will be understood from the introductory discussion of castrate atrophy (at the beginning of this chapter), these same effects of estrogens occur in adult animals that have been deprived of their ovaries, just as in immature animals. In short, the action of estrogens is to bring up the immature uterus to the full adult stage, and then to keep it up, ready and waiting for the further changes that will be imposed upon it by the action of the corpus luteum. All this has been amply confirmed in the highest animals; in monkeys by direct experiment on immature and castrated animals, in human patients by therapeutic treatment of women whose ovaries had been removed for surgical reasons.

We once saw a fantastic outcome of estrogen treatment. Thomas R. Forbes, while a student in my laboratory at the University of Rochester, tried estrone on some 18-inch female alligators. These creatures were still several years short of sexual maturity, and their oviducts (they have no uteri, being oviparous) were very immature. The heavy dosage of estrone crowded years of development of the oviducts into a few weeks, while the rest of the alligator's body remained small. Before long their bellies began to bulge, they sickened and died. At post-mortem Forbes found the abdominal cavities full of nothing else but coil after coil of hypertrophied oviducts, which had pushed the liver up toward the head, jammed the intestines into the flanks, and finally killed the unfortunate creatures when they could no longer find room for these redundant viscera. It would not be possible (I hasten to add) to produce such tragically disproportionate growth in humans or other mammals except perhaps by treatment during the embryonic period.

In mammals, of course, the vagina shares in all this response of the reproductive system to estrogenic hormones, and in the rat, mouse and guinea pig it gives tell-tale signs of its response by such changes in the vaginal wall and in the free vaginal cells as we have already discussed and illustrated (page 83; Plate XV). In human females who have passed the menopause, and younger women whose ovaries have been removed surgically, the vagina shows the effect of diminished or absent estrogenic hormone. When such patients are given the hormone, its effect can be detected by studying the vaginal cells. This method is now used to diagnose loss of ovarian function and to follow the progress of treatment with hormones.

The estrogenic hormones act on the mammary glands in a very definite way. These glands, when fully developed, are constructed on a plan resembling a little cluster of trees, in which the trunks come together at the nipple, forming the main milk canals. The branches form the general duct system, and the leaves are the milk-producing units (see Fig. 29). In immature animals the twig-like terminal units are scarcely or not at all developed, and even the duct system is rudimentary. Estrogenic hormones make the ducts grow until the gland is a wide-spreading tree, but still without much development of the actual milk-producing terminations on the ends of the twigs. In short, the estrogens bring the infantile mammary glands up to the normal stage of development of a mature virgin animal. As we shall see in Chapter VIII, during pregnancy other hormones, namely the pituitary hormones and in some animals the corpus luteum hormone, carry on the development of the gland and bring about the production of milk.

The estrogens and puberty. When the time approaches for a young girl to become mature, the pituitary gland begins to put out its gonadotrophic hormones (see p. 141). These in turn stimulate the ovary to produce its estrogenic hormone. All the accessory reproductive organs (uterus, oviducts, vagina, external genital organs, mammary glands) begin to grow toward mature size. The menstrual cycle begins. Hair grows in the armpits and the external genital region. All these changes are due directly to the estrogenic hormone. If the ovaries are removed before puberty they do not occur, and on the other hand they can be brought out in experimental animals and in castrated women by suitable injections of the hormone. Other signs of mature femininity, namely the female type of skeleton and bodily contours, the adult type of voice, are more deeply innate characteristics. It would be a mistake to think that the estrogenic hormone and the male hormone make all the difference of bodily type between men and women, or between male and female animals. Sex is determined when the egg is fertilized.[3] The animal develops ovaries because she is a female already. When they begin to function as endocrine organs they make her sex effective by developing the accessory sexual organs so she can ultimately bear her young. In this sense the estrogenic hormone is a sex hormone; but if the ovaries fail to develop or are removed in childhood, and the ovarian hormones are thus unavailable, the girl still becomes a woman—infertile of course, usually somewhat immature or boyish, but still physically a woman, not a male or a neutral individual. For this reason the term "female sex hormone" has been generally abandoned and the safer name, estrogenic hormone, used instead (Appendix II, note 3).

Estrogens and the estrous cycle. Given a castrated female guinea pig and a syringe of estrogenic hormone, the experimenter can reproduce cyclic changes in the vagina exactly like those found at estrus in the normal animal whose ovarian timepiece is ticking properly; and if he times his injections carefully he can set up a regular vaginal rhythm every 15 days (the normal rate of this species), so that an observer following the cycle by examining the vaginal cells, could not discover the absence of the animal's own ovaries.

This artificial cycle will also affect the uterus. The fort-

[3] See any of the books on heredity cited in note 2 of Chapter I.

nightly injections of estrogenic hormone will cause it to undergo the changes characteristic of estrus. The ovaries being absent, there will be no follicles to ripen and hence no corpus luteum. If the experimenter wants to make his artificial cycle complete, he will have to follow up his estrogen with injections of the corpus luteum hormone—but that is next chapter's story.

I have chosen the guinea pig as my example, but I might well have spoken of other animals. In the spayed female dog, for example, all the physical signs of "heat" characteristic of that species are brought on by estrogenic hormones, including the swelling of the external genital region and the sanguineous discharge normally seen at estrus or shortly before. In short, we can say that the physical changes of uterus and vagina that go with the ripening follicle of the ovary, are caused by the estrogenic hormone. In the life of the normal animal, however, these changes are accompanied also, during the estrous period, by the psychological urge to mate. Is this also produced by the estrogenic hormone of the ovary? Not all the necessary evidence on this important question is at hand. Psychic reactions are notoriously difficult to study. It is not easy to observe, test, and measure the sex behavior of animals in the laboratory. We know of course that if the ovaries are removed, the female is no longer receptive sexually, but even this statement requires reservations, especially in the case of our own unpredictable species. In many animals that have been studied (e.g. rat, mouse, dog, cat, sheep) there is evidence that the administration of estrogenic hormone to castrated females, or to intact females outside of the mating season, will awaken sexual receptivity. Josephine Ball found the same result in the Rhesus monkeys in the colony of the Carnegie Embryological Laboratory. In the guinea pig, recent work by William Young and collaborators at Brown University and Yale Medical School strongly suggests that not only the estrogenic hormone but also that of the corpus luteum is necessary. At any rate, these investigators obtained more frequent and more normal responses in castrated female guinea pigs if they added a little corpus luteum hormone after a course of the estrogen. This sets up a neat dilemma: there is of course no corpus luteum, in the ordinary cycle of normal animals, until after the animal has been in estrus. We can only conjecture that in the guinea pig at least, the mature follicles must secrete a little corpus luteum hormone before they rupture and become actual corpora lutea.

Important as this question of the relation between estrogens and the estrous urge is, we simply do not know enough as yet to apply the above results to human beings. To what extent the sex urge in women is controlled directly by the estrogenic hormone is a question too complex for analysis at present. The behavior of rats and guinea pigs is difficult enough to understand. Human behavior involves all sorts of mental processes not subject to experimental control. We may be sure, however, that the hormones have an important part in the matter, directly or indirectly, and that without them there could be no human mating.

The estrogenic hormone as a medicinal drug. When a physician administers an estrogenic hormone he is giving the patient a substitute for her own hormone. He thinks her supply of estrogen is too low for current needs. This lack is, however, not easy to determine or to measure. The signs of hormone deficiency are not very clear, and laboratory tests by examination of the patient's blood, to see how much hormone it is carrying, are expensive and in our present state of knowledge not necessarily decisive. Upsets of the menstrual cycle and other disorders of the female reproductive system may or may not be due to hormone lack; the theory of menstrual disorders is not yet clearly worked out. For this reason their treatment with hormones is at present a matter for very cautious study by specialists who have the facilities to treat their patients as laboratory subjects, which means more attention, more examination and more testing than patients with better-understood diseases usually require, and more expense for somebody, either the patient herself or the research budget of a clinic. The pages of medical journals are, however, nowadays full of reports of research, and day by day the subject is being advanced.

There is one disorder of health in which it is perfectly clear that the trouble is caused by the ovary failing to put out enough hormone and that a supply from the drug store will be helpful. This is the distress that often follows the cessation of the menstrual cycle, both the natural menopause of women at 45 or 50 years of age, and the so-called "surgical menopause" produced by operative removal of the ovaries. When the characteristic symptoms of a stormy menopause, such as hot flushes of the skin and general nervousness, become almost intolerable, large doses of estrogenic hormones often give genuine relief.

Another helpful use of estrogenic hormones is in the treatment of retarded sex development caused by ovarian insufficiency. In certain cases, selected by experienced physicians as likely to be helped, and carefully treated over long periods, the aspects of femininity have been given to underdeveloped girls, with great benefit to their social adjustment and their morale.

These hormones are powerful agents. We have seen that they affect many organs, and even yet we do not know all they may be doing in the body. In careful experienced medical hands they can be beneficent drugs. Used recklessly they may do great damage.

The estrogenic hormones and cancer. There is a whole series of chemical substances that cause cancer or other tumors, malignant and benign, when injected or rubbed into the skin. Some of these are chemically similar to the estrogenic hormones and indeed there are a few synthetic cancer-producing (carcinogenic) substances that are also estrogenic. It becomes therefore a burning question, whether or not the familiar and commonly known estrogens are carcinogenic. This question cannot be answered "no" or "yes" without explanation. Briefly the situation is as follows: The estrogens commonly used in medical practice and in the laboratory do not ordinarily produce cancer by their own direct action. In experimental animals cancer may follow their use under special circumstances, of which the following is a good example. In a certain pure-bred strain of mice, the females have a high tendency to develop cancer of the mammary glands. The males inherit the same tendency but do not suffer from it because the male mammary gland is too scanty to become cancerous. If the males are given large doses of estrogenic hormone, their mammary glands grow larger and then they often develop cancer of the breast.

In our Carnegie Embryological Laboratory, Hartman and Geschickter have given sixteen Rhesus monkeys enormous doses of estrogenic hormones over periods of many months, even for two or three years, and have not found a single tumor. On the other hand Alexander Lipschütz of Santiago, Chile, with Rigoberto Iglesias, Luis Vargas Fernandez and others has found that in guinea pigs it is very easy to produce fibrous tumors in the abdominal cavity with estrogens. These are not malignant but may kill the animal by their mere size. There is even a recent report (from Gardner and E. Allen of Yale) of tumors of the uterus in mice, which may be malignant, produced by injection of estrogenic hormones in non-cancerous pure-bred strains.

The present definite medical consensus is that in human beings cancer is not produced by ordinary doses of estrogens. The whole subject, however, demands and is getting further investigation.

A HORMONE FOR GESTATION

"In general estrogen is the hormone of the woman, *it assures the development of the genital and mammary apparatus; progesterone is the hormone of the* mother, *it is indispensable for reproduction."*—ROBERT COURRIER, *Biologie des Hormones Sexuelles Femelles,* 1937 (translation).

❦ CHAPTER V ❧

A HORMONE FOR GESTATION

IF this were a detective story, not merely a story of detection, the reader would at this point be directed to turn back to pages 41-44, in which (he would solemnly be informed) he will find all the clues necessary to solve the great Corpus Luteum Mystery. Having reread those pages and having inspected the photographs of the corpus luteum, Plate IX, he will be in possession of all the information that enabled Louis-Auguste Prenant of Nancy in 1898 to suggest that the corpus luteum, subject of so much previous conjecture and so little fact, is actually an organ of internal secretion. This guess, I admit, required sharp wits. Today every reader who has studied biology ought to get the same idea when he sees this small but vivid organ, walled off from the rest of the ovary by its fibrous capsule, drained by no secretory duct but obviously equipped with a rich network of blood vessels, and composed of large and characteristic cells resembling those of the adrenal gland. Seen under the microscope as pictured in Plate IX, *C*, this arrangement fairly shouts "I am a gland of internal secretion"; to recognize that fact forty-four years ago was a real feat of scientific detection.

But even in novels of crime the sleuth's clever guess must be followed by careful accumulation of evidence for the jury; still more is this the case when the object of detection is the function of an important gland and the seeker's reward a valuable addition to scientific knowledge and human welfare.

I do not pretend to write this chapter in cool detachment. Its theme-word *progesterone* has for me connotations that will never be found in the dictionary. In the first place I invented the word myself, as far at least as the letter "t," as will be explained hereafter. In the second place, it recalls memories of bafflement, comedy, hard work, and modest success. Can I forget the time I went racing up the steps of the laboratory in Rochester, carrying a glass syringe that contained the world's entire supply of crude progesterone, stumbled and fell and lost it all? Or the day Willard Allen showed me his first glittering crystals of the hormone, chemically pure at last?

In the third place, I am writing largely about the work of personal friends. Prenant and Born I did not know, for chronological reasons; but Paul Bouin received me in his laboratory at Strasbourg many times in the summer of 1924, looking like a Frenchman out of a storybook, writing and teaching like the grand scholar and gentleman he is. The story of Born's legacy I heard from Ludwig Fraenkel himself, to whom in Montevideo may this book carry a special greeting. Dispossessed of his famous clinic in Breslau, driven from his country, he can never be exiled from a world that honors great minds and great hearts wherever they are. Karl Slotta and Eric Fels, when at last I met them in their South American homes, proved no less distinguished in hospitality than in chemistry. Adolf Butenandt of Berlin sat happily at my own fireside and dinner table in 1935, and there he will be welcome again now that his country's guns are silenced.

Nor can I possibly write with detachment of Willard Allen's work, which at first I shared and afterward watched with affectionate admiration. The American pioneers in this work, Leo Loeb and Robert T. Frank, early honored me with their acquaintance and good will. Likewise the names of our Wisconsin fellow-workers, Frederick Hisaw (now at Harvard), Harry Fevold, Charles Weichert, Samuel Leonard, Roland Meyer (the latter for three years also in my laboratory at Rochester) are written not only in the formal list of investigators to be mentioned here, but also in memory's record of friendly rivalry and mutual enthusiasm. Without apology, then, let these personal feelings color (for so they must) the narrative of research.

The collection of evidence begins with a scene poignant enough, indeed, for a novel. In 1900 the great embryologist of Breslau, Gustav Born, lay dying. Scientist to the last, his mind was full of a hypothesis he knew he could not live to test and which he could not bear to leave untried. To his bedside, therefore, he summoned one of his former students, the rising young gynecologist Ludwig Fraenkel. To him Born imparted his thought that the corpus luteum is indeed an organ of internal secretion, and moreover that its function must be concerned with the protection of the early embryo. This guess about its specific function, like Prenant's about its general nature, was brilliant and novel in its day, even though to us in retrospect when we consider that the corpus luteum is present only when the egg is available for development, such a function seems probable indeed. So it seemed then to Fraenkel, whose task it was to devise the experiments by which Born's conjecture could be put to rigorous test.

He knew that in the rabbit the embryos become implanted in the uterus on the 8th day after mating. They spend 3 days in the oviduct and then 4 days more as free blastocysts in the uterus, before they become attached. To test the function of the corpora lutea, Fraenkel planned to interfere with the natural course of events by removing them while the embryos were still unattached. The simplest way of removing the corpora lutea is of course to remove both ovaries, by surgical operation under an anesthetic. Since this might, for all he knew, remove some other useful or necessary factor, Fraenkel tried also cutting out the corpora lutea alone, or burning them out with a fine cautery, of course always under an anesthetic. This operation is more difficult than simple removal of the ovaries. The rabbit sheds many eggs at a time, up to 10 or even 12, and a corresponding number of corpora lutea have to be searched for and removed at the operation.

As already explained on p. 68, rabbits are peculiar in that the ovarian follicles ripen only after mating, not spontaneously at more or less regular intervals, as in other animals. This is a great advantage for the present purpose, for it means that we can time our experiments at will. Fraenkel simply mated his females to buck rabbits of known fertility. He knew they would ovulate next day and that while the eggs were being fertilized and beginning to develop, the ruptured follicles would be transformed into corpora lutea. Sometime during the next 6 days he intervened surgically and destroyed the corpora lutea. Then he simply waited to see what happened. If loss of the corpora did nothing, he could expect that after 3 weeks the rabbit would show signs of pregnancy and about the 33rd day, as is the rule in this species, she would give birth to her litter of young. Actually, when the experiment was performed as described, no pregnancy ensued. Something had interfered with the embryos. Fraenkel checked this result by careful control experiments; he removed only one ovary, or cut into both ovaries without removing the corpora lutea. Thus he had experiments in which there was no endocrine loss, but just as much upset and damage as if the corpora lutea had been removed. The results were decisive. If the corpora lutea were not completely removed, the pregnancy went on. If they were removed, the pregnancy failed. In many cases he did not wait for the time of birth, but autopsied the rabbit 3 or 4 weeks after operation, always finding that the embryos had disappeared from the uterus. He did not learn what had actually happened to them, nor when the blow fell. He only knew they could not survive the loss of the corpora lutea.

These results were presented to the German Gynecological Society in 1903, but they met a good deal of criticism and disbelief. Some of the experiments were for technical reasons not perfectly convincing, although we know now their outcome was correct. Fraenkel returned to his laboratory and after seven years was able to present in 1910 completely acceptable results.

At about the same time as these later experiments, and apparently without knowledge of Fraenkel's work, the French histologist Bouin and his colleague Ancel had under way experiments (published in 1910) which demonstrated another aspect of the activity of the corpus luteum. They found that during early pregnancy the lining of the uterus undergoes a remarkable change. This is shown in Plate XVII, *B*, which reproduces one of Ancel and Bouin's actual figures.[1] The tubular glands of the uterus begin to grow longer, to secrete fluid and therefore to become dilated. Their cells multiply so fast that there is no longer enough room for them in the simple tubular wall, and the glands begin to become folded or pleated. The folds of the endometrium are deeply pervaded by these glands; and finally in a section of the uterus one sees a beautiful lacelike pattern (Plate XVII, *B*, right) representing the cross-section of this gland-filled tissue.

This change in the condition of the lining of the uterus, described by Ancel and Bouin, gave the key to the discovery of the corpus luteum hormone. We shall have to mention it again and again. We can therefore save words by using the technical name of this change, i.e. *progestational proliferation*, that is to say "growth and change which favors gestation." Obviously, as the picture shows, it *is* growth and change; that it favors gestation will be proved as we go along.

Anticipating our story again, so that we may be perfectly clear about this important matter, let it be said that progestational proliferation does not occur in the rabbit only, but

[1] Photographs of the rabbit's uterus magnified, shown in this book, represent sections (slices) across the uterus, as one slices a banana.

in all mammals, in the first days of pregnancy and also without pregnancy, whenever a recent corpus luteum is present in the ovary. This means that in animals with regular cyclic ovulation, for example the guinea pig, dog, cat, pig, human, and indeed most mammals, progestational proliferation occurs in each cycle. It is not exactly alike in all animals; it is very elaborate in some, such as the rabbit and the primates (monkeys, apes, and humans) but in others, e.g. rat, mouse, and guinea pig, it is relatively slight. Plate XVII shows it in three kinds of animals.

Ancel and Bouin guessed that this progestational proliferation is caused by the corpus luteum, and they took steps to prove it by an ingenious plan. They mated their female rabbits, not to fertile bucks, but to males rendered infertile by tying off their seminal ducts. Such rabbits ovulated and formed corpora lutea but, of course, did not become pregnant. The only respect in which they were different after mating was that their ovaries now contained corpora lutea. Since their uteri developed typical progestational proliferation, the corpora lutea must have been responsible. The two experimenters then repeated their experiment of mating their rabbits to infertile males, but within a day or two they removed the ovaries or cut out the corpora lutea. Progestational proliferation did not occur. Obviously the corpus luteum controls the condition of the

PLATE XVII. Preparation of the uterus for implantation of the embryo (progestational proliferation) in human, rabbit, and pig. In each case the left-hand figure shows the interval stage, the right-hand figure shows the effect of the corpus luteum hormone. *A*, this process in the human uterus, from the first description by Hitschmann and Adler, 1908. Magnified about 15 times. *B*, the first pictures of progestational proliferation of the rabbit's uterus, by Bouin and Ancel, 1910. Magnified about 10 times. *C*, the same change in the uterus of the sow, from preparations by the author. The left-hand figure represents the day before ovulation, the right-hand section was taken 8 days after ovulation. Magnified 10 times.

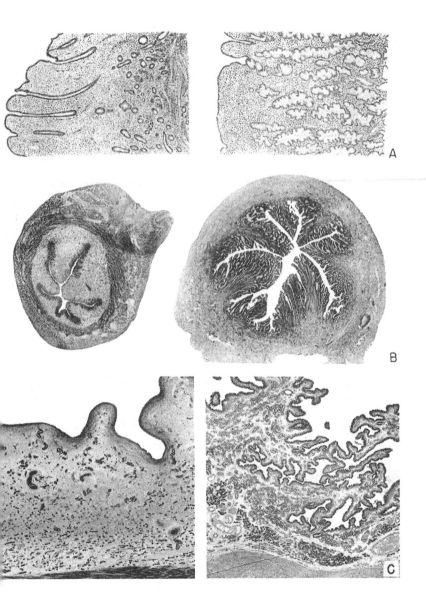

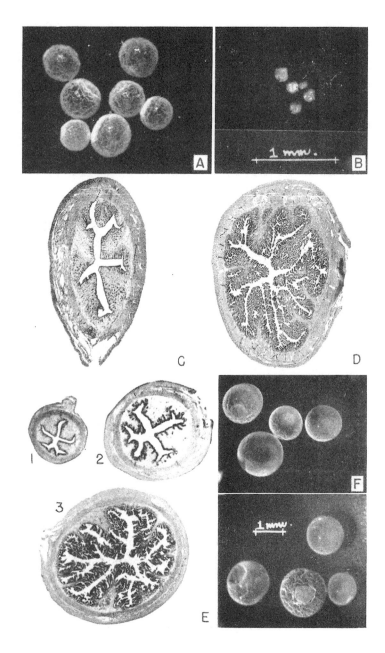

uterus and determines the occurrence of progestational proliferation.

The American investigator Leo Loeb had indeed already shown (1909), in another way, that the corpus luteum controls the state of the uterus. In the guinea pig the embryo settles deeply into the uterine lining (endometrium) as it does in the human species, and the maternal tissue responds by active growth at the site of attachment, so that the placenta contains a great mass of maternal cells. Loeb showed that the maternal response is caused to occur by the irritation, so to speak, produced by the embryo as it settles into the lining of the uterus. In his experiments he imitated this irritative action, in nonpregnant animals, by putting into the uterus, not embryos, but bits of foreign material, such as tiny pieces of glass or collodion. In his simplest trials he ran a silk thread through the uterus and tied it in place, or merely inserted a needle into the uterus and scratched the endometrium. At the points of irritation, the interior of the uterus promptly developed within a few days, little tumors made up of cells closely resembling, under the microscope, the maternal part of the placenta.

Now we come to the point of all this. Loeb discovered that he could get his tumors only during a limited part of each cycle, when the corpora lutea are present. If he tried at other times, or if he tried it at the right time but took away the

PLATE XVIII. Action of progesterone, the hormone of the corpus luteum. A, normal litter of embryos of rabbit in uterus, 5 days after mating. B, dead and degenerating embryos (same age as those in A) in uterus of rabbit whose ovaries were removed the day after mating. Magnified 20 times. C, section of uterus at time of ovulation. D, uterus of rabbit castrated just after ovulation and given injections of progesterone for 5 days. Magnified 7 times. E, at (1) section of uterus of infant rabbit 8 weeks old; at (2) same after 5 days' treatment with estrogenic hormone; at (3) same after 5 more days' treatment with progesterone. All magnified 7 times. F, two litters of rabbit embryos 5 days old, from mothers deprived of their ovaries one day after mating but injected with progesterone daily. Magnified 7½ times. Preparations by author. Courtesy American Journal of Physiology.

corpora lutea, then he got no placenta-like tumors. In short, when the corpora lutea are present, the uterus is in a special state in which it can respond to the need of the embryos for maternal protection.

Reviewing the story, we see that Fraenkel demonstrated that the implantation of the embryos depends upon the corpus luteum, while Ancel and Bouin had shown by one experiment, and Loeb by another, that the functional state of the uterus depends upon this same gland. We cannot escape the conclusion that these two facts are connected; in other words, that the corpus luteum fosters the embryos by setting up progestational proliferation. No doubt this would have been proved very promptly had not the World War of 1914-1918 interfered with such investigation.

It fell to my own lot (in 1928) to conduct the experiments which tied together the discoveries of Fraenkel and of Bouin and Ancel, by showing exactly what happens, and when, to embryos deprived of the support normally afforded them by the corpus luteum. In the first of my experiments I mated seven female rabbits to fertile males. Fourteen to eighteen hours thereafter the ovaries were removed by surgical operation under complete anesthesia. At this time, as we know from previous studies by embryologists, the eggs were in the 2-cell stage and were in the oviducts. Five, six, or seven days thereafter the animals were killed and examined. Progestational proliferation had of course not occurred, because the corpora lutea had been removed. When the embryos were recovered, by washing them out of the uterus, it was found that they had died *in utero*. From their measurements and stage of development, as compared with normal embryos (Plate XVIII, *B*, *A*), it could be ascertained that they had ceased to grow on the fourth day, i.e. as soon as they had entered the uterus. These embryos died because the uterus was unprepared to receive them.

A control experiment was done with seven more rabbits.

In these everything was done as before, except the ovaries were not removed. Parts of them were removed, or they were cut in two, leaving their blood supply intact; in short, as much interference and damage was produced as in the first group, but in each case several corpora lutea were left in place. In these rabbits, progestational proliferation of the uterus occurred normally; and in six of the seven, normal embryos were found when the rabbits were killed for study on the 5th to the 8th day.

To sharpen the results, and pin them directly to the corpus luteum, I was luckily able to find seven rabbits in which, when I explored them, the corpora lutea were found to be grouped all together in one end of an ovary. When this chanced to be the case, in either the right or the left ovary, I could take out one ovary entirely and all the grouped corpora lutea from the other, still leaving a large amount of ovarian tissue. In all these, progestational proliferation failed to occur and the embryos died.

We have proved two points. First, we have shown that successful care of the embryos *in utero* depends upon a chain of events. The corpora lutea prepare the uterus, the uterus then cares for the embryos. The reason, unknown to Fraenkel, that his embryos died was that he had prevented Ancel and Bouin's progestational proliferation, by removing the corpora lutea.

Second, we have found that the corpora lutea are necessary not only for implantation (as indicated by Loeb's experiment), but also still earlier, for the nutrition and protection of the embryos during the time when they are lying free in the uterus. How they do this is another question which will be considered later; obviously it is a matter of chemical substances secreted by the glands of the uterine lining under the influence of the corpus luteum (Appendix II, note 4).

If the corpus luteum can do these remarkable things by hormone action, we ought to be able to get the hormone out of the gland and purify it; but how are we to know when we have it and how much we have? This question is answered by our experiments on rabbits, just cited. All we need to do is to mate a rabbit, remove the ovaries next day, and then administer our extract to see whether it will cause progestational proliferation, and if so, how much extract is required.

THE HORMONE OF THE CORPUS LUTEUM

From this stage on I was fortunate indeed in having the collaboration of Willard M. Allen, then a medical student[2] equipped with an excellent knowledge of organic chemistry. We began, of course, in the dark. All we knew for certain was that we had to extract something; we did not know what it was or what its chemical properties might be. We had two clues. In the first place, practically all the known important chemical substances in the animal body can be dissolved and therefore extracted by either water or alcohol, provided they are protected from breaking down, spoiling, or being digested in the process. (Incidentally, how can you protect a substance from spoiling if you do not know what it is?) In the second place, we had a hint from the work of Edmund Herrmann, mentioned in Chapter IV, p. 82. Some of his photographs, published in 1915 in the report of his work on the ovarian hormone (i.e. estrogen) showed that without realizing it he had produced progestational proliferation with some of his extracts. From his report we knew that whatever he had in his extracts must be soluble in alcohol. Willard Allen and I began therefore by collecting corpora lutea of sows' ovaries from the slaughterhouse. We minced them up in a meat chopper and extracted them with hot alcohol. Very luckily for us, Walter R. Bloor, then professor of biochemistry at Rochester, is a great expert on animal fats and allied substances. We built our extractors from his design and sought his advice on

[2] Now Professor of Obstetrics and Gynecology, Washington University, and head of the St. Louis Maternity Hospital.

many details of the chemical manipulation. After various tribulations, one of which I shall narrate below (page 118), we found (1929) that we could obtain a crude oily extract, looking like a poor grade of automobile grease, which when injected into experimental rabbits was a perfect substitute for their own ovaries in tests such as described above. After 5 days' injection, progestational proliferation was complete. This is well illustrated by comparison of Figures C and D on Plate XVIII, which show sections of the uterus of a castrated rabbit before and after treatment.

In some of our experiments, in which a large dose was used, we even improved upon nature by producing more extensive progestational proliferation than normally occurs. What was even more striking, the embryos at 6 days were just as well off as if their mother's own ovaries had seen to their welfare. Plate XVIII, F shows two such litters of embryos 6 days old, in rabbits castrated more than 5 days before. Evidently our crude oily extract contained the long-sought hormone.

At the University of Wisconsin, at the same time, F. L. Hisaw and his associates, Meyer, Weichert and Fevold, were also engaged in the extraction of sows' ovaries in the search for an active substance. Their reason for looking for a hormone was, however, curiously different. In the pregnant guinea pig, as the time of birth approaches, the two pelvic bones, right and left, become separated where they join in front; that is, on the belly side of the animal. The bony junction in fact almost melts away, in order to allow birth of the infant guinea pigs, which are so large in proportion to the mother that they cannot pass through the pelvis as it normally exists. This is one of the strangest of those strange adaptations which make the reproductive functions of one species different from those of another, and which puzzle and confuse the investigator, but sometimes offer unexpected clues if he is sharp enough to see them.

Looking for the cause of this pelvic relaxation in the

guinea pig, Hisaw and his young fellow students sought for a hormone ("relaxin") in the corpus luteum. In the course of this quest they obtained, and were the first to mention (1928), although without exact definition, extracts having some of the properties now known to be those of the corpus luteum hormone. The matter of the supposed relaxative hormone, incidentally, still remains a puzzle about which too little is known to discuss it in this book.

With one of these preparations Weichert (1928) was able to duplicate Leo Loeb's experiment (described above) in a castrated guinea pig; that is, the corpus luteum extract acted upon the guinea pig's uterus so that in response to irritation of its lining it produced masses of maternal cells as in pregnancy. This was confirmed by several of my students. Thus the basic functions of the corpus luteum suspected by Ancel and Bouin and by Loeb were confirmed by the use of extracts. Fraenkel's findings were also soon repeated with the hormone, for as soon as Willard Allen and I could prepare enough of the extract, we were able (1930) to castrate female rabbits at the eighteenth hour after mating and to carry the embryos, by use of our extract, all the way to normal birth at the usual term of pregnancy, in seven of our first fourteen attempts.

It should be added that this is a tricky experiment about which even yet we do not have full information. It seems to require just the right amount of estrogen along with the progesterone. Perhaps we were lucky that our extract was just sufficiently impure. At any rate it has turned out to be much harder to accomplish the same result with chemically pure progesterone.

Although the experiments thus far cited were all done upon the rabbit and guinea pig, there is no doubt that similar effects are produced in other species, including the human. In 1930 Hisaw and Fevold were able to produce progestational proliferation in the monkey, and the same result has since been achieved many times in women by gynecologists who had occasion to administer progesterone to human patients whose ovaries had been removed. There will be much more to say about this when we come to discuss the menstrual cycle in Chapter VI.

Meanwhile Willard Allen was making successful efforts to purify the hormone. Our crude extracts were already free of protein and we had got rid of the phosphorus-containing fats, always a bother in alcoholic extracts. The next stages were much harder than it was to purify estrone, because unfortunately the corpus luteum hormone is destroyed by alkalies. The chemists, therefore, could not get rid of fatty contaminants by merely turning them into soap and washing them away. By one trick and another, however, Allen successively got rid of the major contaminants, namely fats, fatty acids, and the inert sterol known as cholesterol. He began to get the hormone in crystalline form. Fels and Slotta, then (1931) in Breslau, Fevold and Hisaw (1932) and Allen (1932) ultimately obtained an almost perfectly pure crystalline substance of high potency. For his part of this work Willard Allen received the 1935 Eli Lilly Award of the American Chemical Society for the best work in biochemistry done by an American under thirty-one years of age. We now called into consultation Dr. Oskar Wintersteiner of Columbia University, a skilled microanalyst. Allen gave him 75 milligrams of the crystals, not much more than the weight of a postage stamp. He completed the purification and found that the hormone is a sterol having 21 atoms of carbon, 30 of hydrogen and 2 of oxygen. Practically at the same time Slotta, Ruschig and Fels at Breslau, and Butenandt and Westphal, then at Danzig, also reported ultimate purification of the hormone and announced, as probably correct, the formula printed below. Butenandt told me later that the total amount of the hormone he had available with which to work out its structure was 20 milligrams, one-third of the weight of a postage stamp. The same year (1934) Butenandt and two colleagues succeeded in making the hormone synthetically by chemical manipulation and rearrangement of a better known and more widely available sterol and thus confirmed the formula.

The chemistry of progesterone. This hormone is also a sterol, and is put together in a way not greatly unlike the estrogenic substances. Students of organic chemistry will recognize it as 3, 20 diketo 4, 5 pregnene:

PROGESTERONE

A more detailed explanation of its chemical relationships will be found in Appendix I. This substance has not yet been synthesized from simple materials, but it has been made by rearranging the structure of somewhat more complicated sterols built up by plants and animals. For some years a vegetable sterol from soy beans was the most readily available source for the synthetic chemist, but it is now being made from cholesterol, which occurs plentifully in the spinal cord of oxen.

The name progesterone. It finally became necessary to name this substance, even before we knew what it was, in order to avoid long phrases in talking about it. Our experiments had proved that its effects are progestational, i.e. it favors gestation; for this reason I decided, with Willard Allen's approval, to call it *progestin.* This word is easy to spell and pronounce in many tongues, means something but not too much, and did not commit us to any theories that might prove untenable later. When the exact chemical nature of the hormone became known, the chemists, led by the amiable and diplomatic Butenandt, suggested the suffix *-sterone*. This tells us that the substance is a sterol containing doubly-linked oxygen.

Progesterone it is, then; and my original name *progestin* is retained for use as a general term when we need to talk about such a hormone without specifying its exact chemical structure. As a matter of fact, some of the other sterols have been found to produce progestational proliferation, though rather feebly in comparison with progesterone itself, and as a group we can call these "progestins," just as we use estrogen as a general term and specify estrone, estradiol, etc. as individual substances.

Natural sources of progesterone. The corpus luteum is the only important natural source of progesterone. It has been extracted not only from the sow, but also from the whale, whose 2-pound corpus luteum contains a large amount. A progestin (probably progesterone) has been extracted from the cow. There is a little in the human placenta, and most curiously of all, the adrenal gland contains something that gives similar effects in animals.

Potency of progesterone; administration. This hormone is not as potent, weight for weight, as the estrogens. Direct comparisons are scarcely possible, for the two kinds of substances do different things; but if we compare the amounts necessary to produce definite effects in the whole uterus and whole vagina of an animal respectively, we have to use doses of progesterone several hundred times larger than of estrogen. Like coal and dynamite, they exert their power in different ways.

In 1935 the League of Nations Commission on Biological Standardization agreed upon an international unit of progesterone, namely the amount of potency in 1 milligram of the chemically pure hormone. This is 1/60 of the weight of a postage stamp. To compare an unknown preparation with this, it must be tested on rabbits under standard conditions.

Progesterone is soluble in oils and fats as well as in fat solvents like ether, and therefore it is generally injected hypodermically in a bland vegetable oil, such as sesame oil. It is ineffective when given by mouth, as shown by extensive trials on rabbits. Recently some of the drug manufacturers have put out another substance, a progestin of slightly different chemical structure, which is reported to give progestational proliferation in rabbits when given by mouth. Its usefulness in human patients is now being established (Appendix II, note 5).

A failure and what it taught. Willard Allen and I had a queer experience with our first extracts, from which we learned something important, so that the story is not only amusing but useful. The beginning of this tale is that when we started we followed (as I said before) a hint from the work of Edmund Herrmann, who had obviously produced progestational proliferation in a few of his experiments without knowing it. He had used very young rabbits, roughly 8 weeks old. They react more readily than adults to the estrogen which was the chief ingredient of his extracts. Since we wanted to follow his methods closely at first, we used infant rabbits too, and with them our first successes were obtained. In the spring of 1929 we were all ready to report the first steps in print. The paper was being written, when it occurred to me that our directions for extracting the hormone ought to be tried out by a none-too-good chemist, just to make sure they were foolproof. We did not want others to think our work could not be repeated, just because our directions were not clear. It was agreed that I was a bad enough chemist for the test: if I could make the extract all by myself, then anybody could. So Allen went on his vacation and I went back to our extractors and vacuum stills. In a week I had a batch ready; to my horror it was ineffectual. I made another batch; it, too, was worthless. I suppressed the paper and telegraphed for Allen. We decided that I needed a vacation and that we would look for the trouble in the fall. In September I made another batch with Allen watching every step, but not touching the apparatus. It was no good. What could be wrong? Since my laboratory was sunnier than his, perhaps my hormone was being spoiled by sunlight. I had a room blacked out and made a batch in the dark. That failed. Then we remembered that Allen, being a better chemist than I, usually got his extracts freer of superfluous grease and therefore had to mix them with corn oil (Mazola) so that he could inject them. Mine were greasy enough to inject without added oil. Perhaps the corn oil protected his hormones somehow while mine spoiled. We checked that idea—another two weeks gone—and that was not the answer. Then in desperation we made a batch together, side by side and almost hand in hand, each watching the other. We divided it into two lots and tested it separately—Allen's worked; mine did not! Eureka, my trouble was in the testing, not in the cookery.

The explanation will seem so silly that I almost hesitate to admit what it was. The fact is that rabbits do not respond well to progesterone until they are about 8 weeks old and weigh about 800 grams. We did not know this, and our rabbits ranged from 600 to 1,200 grams. When we went to the cages to inject them, Willard Allen's idea of what constitutes a nice rabbit led him to choose the larger ones, while I must have had a subconscious preference for the infants. My extracts had been as good as his all the while, but my rabbits were insensitive. It is staggering to think how often the success or failure of research may hang upon such an unimaginable contingency.

We thought very hard about this experience, and decided that after all we should not have expected that the hormone of gestation would act upon an infant's uterus. Caring for embryos is a job for a grown-up uterus. Thereafter we used only adult rabbits and never had another failure. Before we published the method one of our laboratory technicians made a lot of the extract unassisted and it worked.

Meanwhile the Wisconsin workers, Hisaw and his colleagues, had found that their extracts that relaxed the guinea pig's pelvic bones would not work in infant guinea pigs unless the animal was first "primed" with estrogenic hormone. Following up this clue, Allen learned that if we make the infant uterus grow larger by a few days' treatment with an estrogen, it will then respond fully to progesterone. A very striking experiment illustrating this is shown in Plate XVIII, *E*. An infant rabbit has, of course, a very small uterus. After 5 days' treatment with estrogenic hormone, the uterus becomes large and well differentiated. If progesterone is then given for 5 days, the uterus shows full progestational proliferation like that of an adult in early pregnancy. Meanwhile the rabbit herself remains a baby small enough to hold on the palm of one's hand.

By what means does progesterone affect the uterus? Just how the various hormones exert their action upon the organs which they control is very obscure. About the action of progesterone we have at present only a very remote idea. We can guess that in some way the hormone alters the general nutritional state of the cells of the uterus by way of a change in their metabolic rate or some other deep-seated effect on cellular physiology.

What happens to progesterone after it acts? In 1936 a discovery of great importance was made by Ethel H. Venning and J. S. V. Browne of Montreal. This is that in the human body used-up progesterone is converted into another substance called pregnanediol (preg-nane-di-ol). The conversion takes place by the addition of six atoms of hydrogen. I give the chemical formula for those who are interested:

PREGNANEDIOL

Pregnanediol is an inert substance as far as hormone action is concerned. The body gets rid of it by attaching it to another substance readily available from the starches and sugars of the food, namely glycuronic acid (see Appendix, p. 253). To this an atom of sodium is added, and the combined substance, sodium pregnanediol glycuronidate, passes out through the kidneys. It happens to be easily separable from the urine and can thereafter be detected and measured by relatively simple laboratory tests. Each molecule of this waste product in the urine means that a molecule of progesterone was available for conversion. If ten milligrams, for instance, of pregnanediol is recoverable from a single day's urine, then we know the patient produced at least that much progesterone—in fact a little more, for there is some loss in the chemical process of measurement and possibly some loss in the body, i.e. some of the progesterone may be broken down and eliminated in other ways. The method is good enough, however, to help very much in estimating the functional activity of the corpus luteum in women and is beginning to be used as a method of diagnosing ovarian deficiencies.

All this information about the conversion and excretion of progesterone unfortunately holds good only for the human species and (apparently) the chimpanzee. It is enough to make a laboratory experimenter tear his hair, when he realizes that in other animals progesterone is excreted in some other way, which nobody has been able to discover. Pregnanediol has not been found in the urine of rabbits and monkeys, nor in any other animal. Nor has any other substance that might be derived from progesterone and from nothing else been found in the urine. Either it escapes from the body, in lower animals than man, in some elusive form; or it is broken down so completely that its chemical fragments are undistinguishable among the simple remnants of bodily chemicals that make up the excretions (Appendix II, note 6).

Action of progesterone on the muscle of the uterus. In 1927 a well-known Austrian scientist, Hermann Knaus, dis-

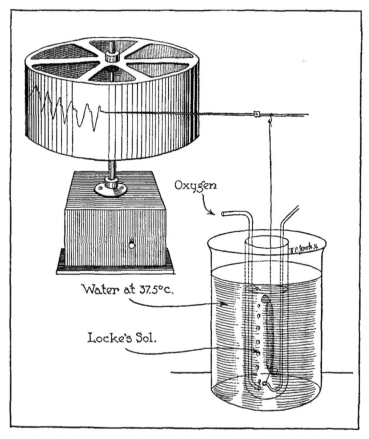

Oxygen

Water at 37.5°C.

Locke's Sol.

Fig. 17. Arrangement of apparatus for maintaining excised pieces of uterine muscle under physiological conditions and recording its contractions.

covered another action of the corpus luteum. To make this clear let us imagine an experiment such as is shown in Fig. 17. A rabbit is killed, the uterus is removed, and a portion of one horn is suspended in a cylinder of salt solution so that one end is tied down and the other is fastened to a writing lever that writes on a revolving drum. The salt solution is kept at body temperature by standing in a bath of warm water. Oxygen is bubbled through the salt solution, and a little sugar is put into it as food, providing energy for the tissues. The whole thickness of the uterus, except its inner lining, is composed of involuntary muscle, as explained on page 48. Placed in an apparatus of the sort just described, which imitates the natural conditions within the body, any such piece of living involuntary muscle, whether from the uterus, intestine, stomach, bladder, blood vessels, or elsewhere, will undergo rhythmic contractions every minute or two and will write them on the revolving drum as shown in the illustration. The experimenter, if he likes, can induce larger, more sudden contractions by putting into the salt solution one or another of those substances that are known to stimulate involuntary muscle. A drop of adrenalin solution will make the uterine muscle pull so strongly that it will almost yank the lever off the drum. So will a drop of pituitrin solution (subject to a very important reservation which is about to be mentioned)—indeed, pituitrin is such a powerful stimulant of uterine muscle that it is often used by obstetricians to make the human uterus contract after childbirth.

Knaus found that the uterus of a pregnant rabbit, set up in salt solution as described, will not react to pituitrin. He had just learned this when the first reports of the isolation of progestin reached Austria; he made up some crude progestin (the first in Europe) and found that when he gave this by injection to a castrated adult rabbit for 5 days, the uterus became absolutely insensitive to pituitrin, exactly as if she were pregnant.

This experiment works with some species of animals and not with others; we do not know surely about the human uterus in this respect. In cats, strangely enough, progesterone suppresses the action of adrenalin on the uterus (something that never happens in the rabbit) but does not suppress the action of pituitrin. These differences present a remarkable and probably very subtle problem in the physiology of muscle. When it is answered we shall know much more than we do now about involuntary muscle and also about hormones. Details aside, however, the experiment of Knaus shows us that progesterone can act in a striking way on involuntary muscle. We owe to S. R. M. Reynolds a very ingenious method of studying uterine contractions in living rabbits. By a simple plastic operation, done in a few minutes under complete anesthesia, the twin cervices of the uterus are stitched to the belly wall and thus made accessible. A rabbit so prepared suffers no inconvenience if properly cared for, and like any other healthy unfrightened rabbit will lie quietly on her back for hours if gently tied down. A tiny rubber balloon is carefully inserted inside the uterus. From this a tube leads to a little bellows which actuates a lever writing upon a revolving smoked drum. Whenever the uterus contracts, the balloon is squeezed, the bellows is inflated, and the lever goes up. With this apparatus Reynolds found that the normal uterus of an adult female rabbit undergoes more or less regular contractions. Castration suppresses the contractions, for the uterus which is enfeebled by castrate atrophy (see page 79) becomes inactive. Administration of estrogenic hormone, however, restores the contractions. Progesterone promptly and effectively quiets the uterus. The graph shown herewith (Fig. 18) illustrates the effect of an injection of progesterone. At 10:20 a.m. the uterus was contracting regularly. The hormone was given a few minutes later. In one hour (see third line of the graph) the contractions became definitely smaller and less frequent.

By 12:20 p.m. the uterine muscle was practically not contracting at all. This effect wears off in a few hours.

Such a sedative action of progesterone upon the uterine muscle has been observed in many animals and there seems to be good evidence that it occurs in the human species. In all probability it serves to keep the uterus quiet in early pregnancy so that the embryos can become safely implanted.

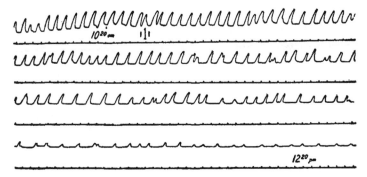

FIG. 18. Effect of progesterone on the contractions of the living rabbit's uterus. A tiny rubber bulb in the cavity of the uterus is compressed each time the uterine muscle squeezes down, so that the rhythmic contractions are recorded on a revolving drum. Progesterone administered at 10:20 a.m. By 12:20 p.m. the uterus is quiescent. From an article by S. R. M. Reynolds and W. M. Allen, 1935, by courtesy of the *American Journal of Obstetrics and Gynecology* and the C. V. Mosby Company.

Is the corpus luteum necessary throughout pregnancy? In the rat and mouse, removal of the ovaries at any time in gestation causes termination of pregnancy. The embryos are cast off prematurely or they break down in the uterus and are absorbed *in situ.* Guinea pigs do not always lose their young if the ovaries are removed after implantation has occurred. In humans it seems certain that both ovaries, including of course the corpus luteum, can be removed after the first few weeks without harm. Whether the pregnant human female, and other animals having long terms of pregnancy, do not need progesterone after the embryos are safely implanted, or whether perhaps the placenta makes enough to serve in place of the corpus luteum, we do not yet know (Appendix II, note 7).

Progesterone as a medicinal drug. Progesterone is already in the drugstores. It comes in neat little boxes of glass ampoules filled with a bright clear oily solution, duly labeled with the hormone content in international units. Tablets of the similar substance that acts by mouth (page 118) are also available for trial. Prescriptions for these drugs will be filled no less readily, though somewhat more expensively, than for digitalis or belladonna.

Before we expect the doctors to cure people with the ovarian hormones, however, let us consider the special circumstances. To take a quite different case, when Banting, Best, Collip, and McLeod handed over the pancreatic hormone, insulin, to the medical profession, they were filling a specific, clearly understood need. They had worked out the insulin problem by experiments on dogs. Sugar is sugar, whether a dog burns it or a man, and the way in which different animals use sugar is the same, regardless of the species. Diabetes, moreover, was a well understood disease, and in the minds of the medical profession it was waiting to be treated with this hormone as a lock waits for the key.

To take another example, the hormone of the adrenal medulla, epinephrin (trade name Adrenalin) has a relatively simple, direct action. What it will do, what it is needed for, is fairly clear. It can be used for acute asthma, or for bleeding from mucous membranes, with understanding and with reasonable hope that it will be effective under the circumstances.

This is unfortunately not the case with endocrine disease of the reproductive system. The human uterus and the rest of the system behave quite differently from those of most other animals. We cannot, for example, apply directly to humans with regard to disturbances of menstruation or pregnancy, information gained from the lower mammals, because the latter do not menstruate and in pregnancy they differ in many ways. It will be clear enough when we deal with the menstrual cycle, in Chapter VI, that we do not even yet fully understand the way the ovarian hormones take part in menstruation. We know that the monthly cycle and the process of gestation require not only the individual hormones but also an exact balance between them. When these things go wrong, they do so in complex and devious ways. When a young woman is cramped with menstrual pain or a wife goes childless against her will, in many cases neither her physician nor the investigator in his laboratory can say exactly what is wrong or how to redress it. They can only try their best; often the treatment works, often not. Only by the slow pathway of experiments on monkeys and cautious observation and trial directly in humans shall we ever comprehend the normal physiology of reproduction in our own species and those vexatious, oftentimes tragic disturbances that lead to disorders of menstruation, miscarriage, and sterility.

It is therefore only in small degree possible as yet to apply progesterone and estrone, and the other potent steroidal hormones, to human disease. We must leave the problem in the hands of competent gynecologists and obstetricians, particularly in the clinics of the medical schools and research hospitals. When these men give the word, hormone treatment becomes justifiable. The work is going forward daily; I do not mean to be discouraging, but only to avoid false promises of quick magic like that of insulin. Incidentally some cases of diabetes still defy insulin, and the specialists in that disease have by no means been able to declare their treatment perfected and their researches complete.

The maintenance of pregnancy. What can we hope for from progesterone in the long run? The great dramatic thing about this hormone as seen in the laboratory is of course its power to maintain pregnancy after the loss of ovarian function. One of the greatest problems of medical practice is that of spontaneous abortion of the embryo or fetus ("miscarriage"). About one pregnancy in three terminates prematurely, according to generally accepted figures.[3] These accidents have many causes. Sometimes the embryo is itself unhealthy, through some mischance of heredity or development, like a seedling plant that will not grow. Sometimes illness or accident to the mother upsets the course of events. Sometimes, we may suppose, the hormones go wrong. Perhaps the supply of progesterone from the corpus luteum or placenta is not adequate. In such cases it might conceivably be useful to supply the hormone by injection, thus making up the lack. The main difficulty at present is to diagnose such cases in time to treat them.

There is another way in which progesterone might help. No matter what the cause of an abortion, it is always accompanied by spasmodic contractions of the uterus, trying to get rid of its burden. Sometimes, we think, the contractions come first, as the result of injury or illness, and dislodge a normal embryo. We have seen that progesterone will quiet the uterine muscle; by reason of this, we may hope that it will help to steady the uterus in the case of a threatening abortion. With these thoughts in mind, the doctor tries progesterone in the face of such a disaster. Sometimes the pregnancy goes on, sometimes not. How can he tell whether his hormone worked the cure? His case records are not altogether helpful, for no two cases are alike; and what is more, the physician in his anxiety generally tries two or three remedial measures at once, any one of which might have been responsible. If he controls a large clinic with many such patients, he can use the treatment on alternate cases only until he is satisfied which group does better. By this method, while the physician

[3] This statement need cause no alarm to any prospective mother who happens to read it. Once pregnancy is past the first weeks and under medical care, it goes safely on in an overwhelming majority of cases. The figure of one loss in three includes many pregnancies of the earliest weeks, and even some that occur so early as to be recognized only by microscopic methods.

is being scientific he is (if the drug is really useful) condemn-
ing half his patients with threatened abortion to the risk of
losing their babies—a dilemma similar to that dramatically
expounded in Sinclair Lewis's great novel of medical life,
Arrowsmith. A clever method of testing this question was
recently suggested by the British physicians Malpas, Mc-
Gregor and Stewart, who pointed out that women who are
unfortunate enough to have three or more successive spontaneous abortions are (statistically speaking) almost certain
to miscarry in the next subsequent pregnancy. If, then, any
kind of treatment is followed by the birth of a living infant,
the odds are great that the medical procedure, and not mere
chance, was responsible. By this severely critical test, it
appears that progesterone is saving some of these babies.
Needless to say, a treatment which is still so largely experimental requires skilled and thoughtful handling, by physicians
thoroughly familiar with the proper dosage and other problems.

Post partum pain. The most clear-cut use of progesterone
involving its sedative action on the uterine muscle is for the
relief of spasmodic pain due to excessive contractions of the
uterus after childbirth. This is sometimes severe enough to
require relief. Lubin and Clarke, of Brooklyn, found that a
single dose of one international unit of progesterone will
relieve these afterpains in about 90 per cent of the cases.

Menstrual cramps. Painful menstruation is one of the commonest of human ills, and one of the least understood. Knowing as little as we do about normal menstruation, it is no
wonder that we also know all too little about its disturbances.
In the case of painful menstruation (dysmenorrhea) we are
not even sure of the exact seat of the pain in all cases. It is
probably due to cramping of the muscular wall of the uterus;
but there is reason to think that in some cases the pain may
be produced in the lining of the uterus rather than in the muscular substance of the wall. All sorts of treatment have been

tried, from psychological analysis to operations designed to correct faulty positions of the uterus. The fact that each of these widely different measures sometimes succeeds and sometimes fails, suggests strongly that dysmenorrhea is not a single disease, but rather a symptom due to different causes in different cases. The whole situation creates a problem for investigation by the combined forces of clinical gynecology and the laboratory investigators. Unfortunately, we can expect to get very little help, in such a problem as this, from study of animals. Even the female Rhesus monkey, so useful for study of the physical aspects of menstruation, cannot help us in this investigation, for even if she suffered from dysmenorrhea or could be made to experience uterine cramps for the purpose of our studies, she cannot report her symptoms or help us evaluate the results of treatment.

When it was discovered that progesterone can quiet the normal contractions of an animal's uterus and even the violent spasms that cause post partum pain in human patients, many physicians thought of trying it in dysmenorrhea, thinking that it might relieve a crampy state of the uterus. As usual whenever a new treatment, no matter what, is tried in one of the old reliable guaranteed-to-baffle diseases, some of the doctors and patients reported hopeful results. A little later skeptical reports began to be published. Critical observers reminded us anew that dysmenorrhea is such a peculiar thing that we must be very cautious about accepting a new cure. For example, there are undoubtedly some cases in which the pain is largely subjective, arising from psychic causes. These people will be helped by any treatment that happens to win their confidence. Depending upon the patient's turn of mind, a hypodermic injection of sterile water given with due assurance, any new hormone in an impressive package, psychological or religious comfort—any of these may give genuine relief. With equal certainty there are other cases produced by some sort of actual physical or chemical disorder in the reproductive system, and these must be attacked, if possible, by treatment aimed directly at the cause. The physician, however, cannot definitely classify these cases before he treats them. Being a merciful man and anxious to give relief as soon as possible, he generally prescribes what has worked best in his last few cases. He usually tries several things at once, thus spoiling a good experiment in the hope of more relief. As a result, it is very difficult to judge the effects of progesterone when used in the treatment of dysmenorrhea. What is badly needed is a large-scale report from one of the university hospital clinics, based on a long series of cases in which progesterone has been used in alternate patients, and in alternate periods in the same patient, so that really scientific checkup of the effects can be provided. Meanwhile, there have been a good many reports of relief of menstrual cramps, some of them almost magical, and other reports of failure. It certainly ought to be tried in cases that have resisted other forms of treatment, and that are severe enough to warrant the necessary hypodermic injections, as well as the expense, which may run up to several dollars at each period if large doses are necessary. If the new progesterone-like drugs for administration by mouth prove to be successful, they will simplify the problem.

Control of irregular or excessive menstrual bleeding. Progesterone has the property of preventing menstruation, as we shall see when we discuss that subject in Chapter VIII. For this reason it is being tried in cases of excessive menstrual bleeding and irregularity. It looks as if this hormone and some of its chemical relatives are going to be really useful in these distressing ailments, as we come to understand them better; but this is decidedly a matter for trained specialists. No drug can safely be used to stop uterine bleeding except after a thorough examination, to rule out the possibility of bleeding from cancer or other tumor of the uterus. Once such causes of bleeding as these are ruled out, the physician may safely try progesterone.

At the present time each case of menstrual disorder is a separate problem and both doctor and patient must realize that hormone treatment is experimental. It brings relief from debilitation and misery to some women even now. As we learn more through cautious trial, more will be helped. Meanwhile those of us who have had something to do with the finding of this hormone wish to see it exploited with care and understanding, not discredited by premature advertising and incautious use (Appendix II, note 8).

THE MENSTRUAL CYCLE

"In this Enquiry indeed, which we are now attempting, no less useful than agreeable, the Wits of almost every Age have toil'd: but as there is hardly any Argument, on which Physicians have wrote more; so is there no one, in which they have given less satisfaction to their Readers. . . . I shall not appear to have employed my Time ill, in endeavoring to set the Nature of the Menses in a clearer Light, than I find it hitherto done by Authors. In which Performance the Reader will find nothing abstruse, nothing far removed from common sense: inasmuch as it has been my only Care to find out the Truth, as much as possibly I could."—JOHN FREIND, *Emmenologia*, translated into English by Thomas Dale, 1729.

THE MENSTRUAL CYCLE

SURELY the process of menstruation is one of the strangest things in all Nature. An important organ—the uterus—serving an indispensable function, is overtaken at regular intervals by a destructive change in the structure of its lining, part of which undergoes dissolution with hemorrhage, and must be reorganized in every monthly cycle. The loss of blood from organic tissues, everywhere else in the animal kingdom a sign of injury, even of danger, is in this one organ the evidence of healthy function. To make the puzzle greater, menstruation is by no means general in the animal kingdom, or even among the mammals. It occurs, indeed, only in the human race, in the anthropoid apes (having been observed in chimpanzees and in the gibbons), in the baboons, and in the Old World monkeys; in short, in a closely related group of primates, one little portion only of the great class of Mammalia. No other animals, in forest, plain, or sea, hiding in dens or grazing the fields, undergo in the course of their cycles any such phase of hemorrhage. It is a paradox indeed that this curious phenomenon of periodic breakdown, seemingly an imperfection, a physiological flaw, is characteristic solely of the females of those very animals we are pleased to think the highest of earth's creatures (Appendix II, note 9).

The periodicity of menstruation. In human females, menstruation recurs at intervals of about 4 weeks. There is a common impression that the cycles are normally quite regular, but any woman who will keep an accurate calendar of her cycles will find a surprising variability.

A recent statistical analysis of thousands of records[1] shows in fact that the commonest average cycle length (the "mode"

[1] Leslie B. Arey, "The degree of normal menstrual irregularity." *American Journal of Obstetrics and Gynecology*, vol. 37, pp. 12-29, 1939.

as statisticians say) in adult European and American women is 28 days, but it is also very common for a woman to average cycles of 25, 26, 27, 29, and 30 days. The individual woman, moreover, often varies several days, in any one cycle, from her own average. To state this in exact terms, so as to make clear just how much variation a woman may consider normal, is rather difficult, for it is a matter of statistics and such things are hard to translate into everyday language. The clearest statement is that of Professor Arey, already quoted, whose words I paraphrase as follows: let a woman keep a record of her cycles for several years, so that she has enough observations to strike an average. Say, for example, that her personal average is 28 days. Arey's figures show that with this information she cannot hope to predict the onset of any given period with accuracy closer than 2.5 days plus or minus, i.e. she may expect it any day between the 25th and 30th after onset of the last period, and even then one-third of all her cycles will depart still more widely from the average length, say another day or two, and sometimes more. The statement that she averages a 28-day cycle has the same kind of meaning as the average price of eggs during the year, which may be very different from the price at any one time, or a baseball player's average of home runs per game. All it means is that the length of her successive cycles will be within a few days, more or less, of 28 days, seldom however coming out at that precise length. In fact a woman whose cycles were perfectly regular to the day, during many months or years, would be a medical curiosity. No such case has ever been reported.

When we come to consider, a little later, the intricate interplay of hormones that goes to produce the cycles, we shall not be surprised that the timing is not perfectly regular. Nor will we be surprised to learn that in young girls, during the first few months or years of menstrual function, while the endocrine mechanisms are becoming adjusted, the cycle length is extremely variable. Arey compiled the records of 100 girls during about two years after the onset of menstruation, and calculated the average cycle length of each during this epoch of their lives. One-third of these girls actually never had a cycle that corresponded exactly to the day with their own personal averages; in other words, every single cycle varied from the arithmetical norm. During the latter part of adolescence there is considerably greater regularity.

The first menstruation most commonly takes place sometime between the ages of 12 and 14 inclusive. The average age at the time of onset, in the white race at least, is $13\frac{1}{2}$ years, but onset at any age from 11 to 16 may be regarded as normal. Delay beyond 16 is a matter for medical investigation.

The normal duration of the menstrual flow may be from one day to one week; the modal duration is 5 days.

Among the infrahuman primates there is only one, the Rhesus monkey, which has been studied in numbers large enough for positive statement. In this species, observed in captivity in the United States and England, the mode, i.e. the most frequent cycle length, is 28 days, as in women. Individual animals have average cycle lengths of 25 to 31 days, and single cycles vary from 14 days up. Rather scanty observations on chimpanzees, baboons, and a few species of monkeys mostly show averages a little longer than 28 days. This statement might or might not hold good if the statistics were more extensive (Appendix II, note 10).

The poetic suggestion quoted at the head of Chapter III, that the reproductive cycles of living things are part of the rhythms of the universe, must not be taken too literally. Menstruation is not regulated by the moon. It happens that the lunar cycle has the same length to the day as the modal human cycle, but we have seen that the human cycle frequently deviates from the mode, and if, for example, the start of the period coincides with the new moon or any other given lunar phase, the odds are it will be off cycle by at least a day or two next month and perhaps completely out of phase next season. I once had 4 females of the Java monkey (Macaca irus) in a cage adjacent to a large group of the closely related Rhesus monkeys. While the latter were running cycles of 28 days' modal length, their Javanese cousins, living under the same moon, were exhibiting a modal cycle length of 35 days. As mentioned in Chapter III, the cycles of other mammals may vary from 5 days to a year in length, and if we consider the birds and insects we finds cycles of one day to 17 years. If the heavenly bodies are to control these rhythms, the cycle of the 17-year locust calls for a hitherto unknown comet!

The idea of a relation between human menstruation and the moon is, however, ancient and widespread. It was, no doubt, suggested by the obvious and inescapable relation between the moon and the tides of the sea. If the moon can control the ebb and flow of great waters, why not also the tides of human life? Perhaps the popular mind has also been influenced by the fact that outbursts of insanity in women sometimes accompany the menstrual cycle; this seems again to link menstruation with the moon, which has long been considered a cause of lunacy. Then there are, of course, certain special cases in nature in which the life of an animal is directly influenced by the moon or the tides (e.g. the palolo, Chapter III). These cases may have helped to foster the notion we are discussing. As lately as 1898 the eminent Swedish physicist Svante Arrhenius thought he had proved the connection of lunar and menstrual cycles by mathematical evidence, but this has been completely disproved, notably by the English physicians Gunn, Jenkin, and Gunn (1937). It is indeed difficult to conceive of any direct participation of the moon in the reproductive cycles of the land-dwelling primates, for if it were really effective we should expect menstruation to occur at the same phase of the moon in all females of a given species, a state of affairs that would have made the social organization of mankind unthinkably different from what it is.

NATURE OF THE MENSTRUAL CYCLE

Events of the cycle in non-menstruating animals. To get a clear understanding of the process of menstruation, it is necessary to understand first what takes place during the cycle in animals which do not menstruate. This has already been discussed in part and illustrated in previous chapters, and is summarized in the diagram herewith (Fig. 19) which represents the typical or generalized cycle of mammals. If we wish to talk about any one species, we shall have to introduce modifications into this scheme, but as it stands it can be used as a basis for understanding them all. In the upper portion, which shows events in the ovary, we see (beginning at the left) the growth and ripening of the follicle. The moment of rupture of the follicle and discharge of the egg gives a convenient point of division, which we may consider as the start of a new cycle. Looking at the third part of the diagram, that indicating sex activity, we see that ovulation occurs during estrus, an arrangement which is adapted to secure fertilization of the egg. Next, the follicle is converted into a corpus luteum. This in turn runs its course, secreting progesterone for about two weeks (in typical species) and then, if the egg is not fertilized, the corpus luteum suddenly begins to degenerate and ceases to secrete its hormone. Thereafter, a new crop of follicles begins to develop. In some animals the new cycle follows at once (e.g. the guinea pig, which has a cycle of only 15 or 16 days); in others several months may elapse, during which the ovaries are relatively dormant (as in dogs and cats) or a whole year, as in many wild animals.

Digression about the cycle in general. We come now to the fundamental question of the female reproductive cycle, namely what causes the alternations of structure and function in the ovary. When the cycle was first discussed, in Chapter III, we could deal with it only as an observed phenomenon of natural history, but we are now in a position to consider

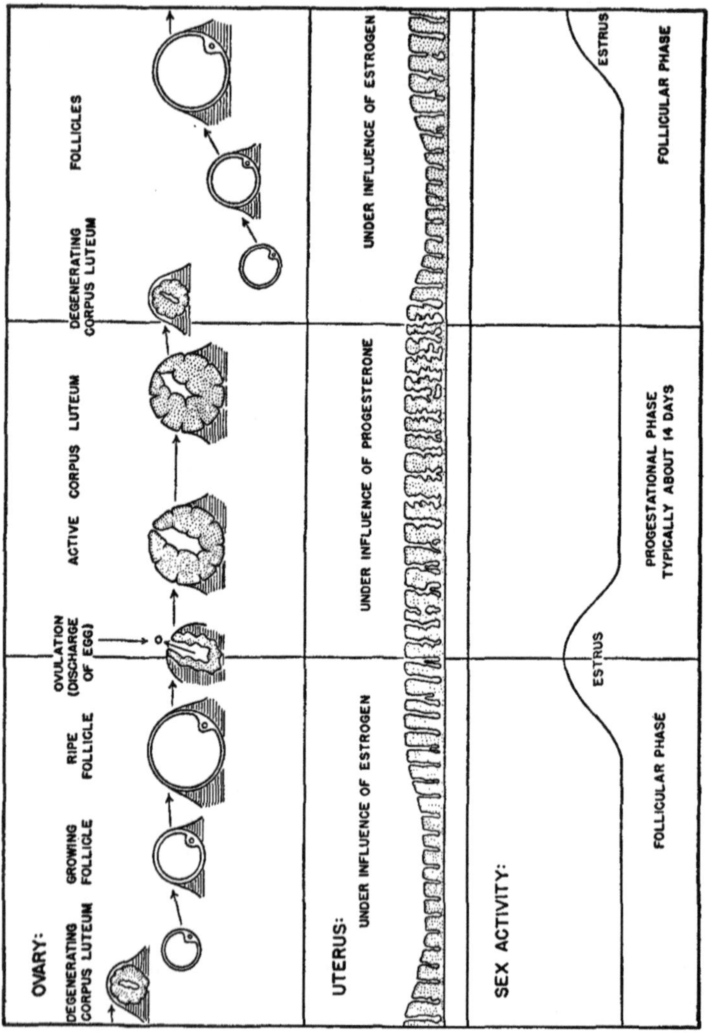

Fig. 19. Diagram illustrating the sequence of events in the typical reproductive cycle of mammals.

the problem in the light of our knowledge of the hormones. To resume this subject where we left off on page 75, there is scarcely any doubt at present that the cycle is somehow produced by interplay of hormones from the ovary and the pituitary gland.

If we look at the pituitary gland or hypophysis (Plate XIX and Fig. 20) we find that this gland of internal secretion is composed of two major parts, the anterior and the posterior lobes. It is the anterior lobe which produces hormones (probably two in number) having the power of stimulating the ovary to produce estrogenic hormones and of promoting the growth of ovarian follicles. They also affect the male organism, causing the testes to grow and produce sperm cells. Because of these actions the hormones we are discussing are called *gonadotrophic*, a name which signifies "producing growth of the sex glands." From the brilliant work of P. E. Smith, Bennett Allen, H. M. Evans, Zondek and Aschheim, and many others between 1915 and the present time, we have learned (as mentioned in Chapter III) that removal of the anterior lobe of the pituitary stops growth of the ovaries and puts an end to the cycles of the animal. By implanting bits of anterior pituitary, or better by injecting extracts of the gland into immature animals, the ovaries are caused to grow and the cycle to begin. The ovary is thus absolutely dependent upon this action of the pituitary. Removal of the anterior lobe produces all the effects of castration, for without it the sex glands, ovary and testis, deteriorate to inactivity. On the other hand, there is a good deal of evidence that the estrogenic hormone of the ovary represses the production of the pituitary gonadotrophic hormones. After removal of the ovaries, the pituitary gland is found to contain more gonadotrophic potency than before; after injection of estrogenic hormones it contains less (Appendix II, note 11).

When these facts became known, a fairly clear explanation of the reproductive cycle suggested itself almost simultane-

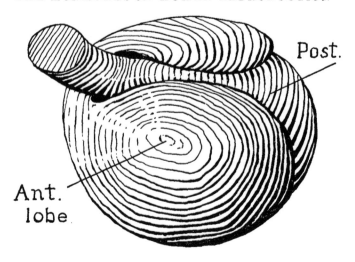

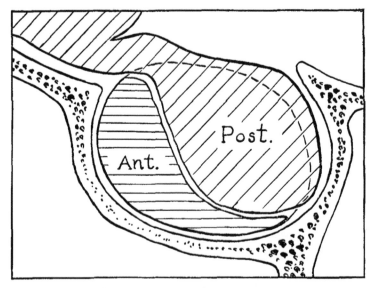

FIG. 20. *Above*, the pituitary gland, showing the anterior lobe and the posterior lobe with its stalk by which the gland is connected to the brain. Enlarged approximately 5 times. *Below*, median section showing position of pituitary gland in its bony cavity at the base of the skull; compare with the X-ray photograph, Plate XIX.

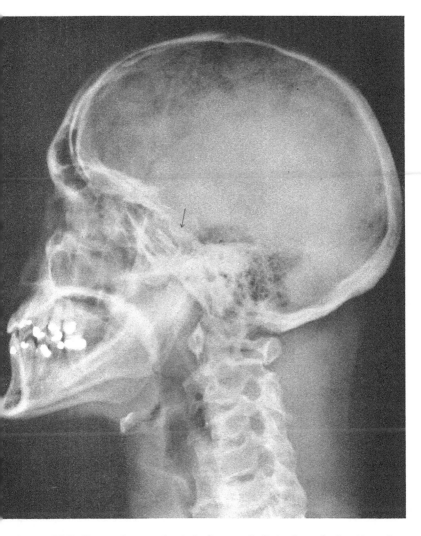

PLATE XIX. X-ray photograph of the human skull, to show the location of the pituitary gland. The arrow points to the little hollow (*sella turcica*) in the bone below the brain, in which the gland lies. About 2/5 natural size. Courtesy of the Eastman Kodak Company, Rochester, N.Y.

PLATE XX. The human infant at birth, with the placenta and membranes. From the anatomical plates of Julius Casserius, published by Adrianus Spigelius in 1626.

ously, about 1931-1932, to a number of investigators, among them first perhaps Brouha and Simonnet in Paris, then to Leonard, Hisaw and Meyer in Wisconsin, and Moore and Price[2] in Chicago. This hypothesis suggests that the cycle is like a clockwork in which the pituitary is the driving force and the regulatory escapement is the reciprocal action of ovarian and pituitary hormones (Fig. 21). The pituitary

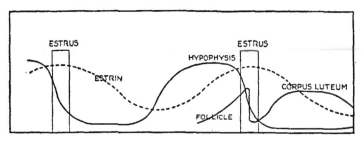

Fig. 21. Diagram illustrating the alternation or "push-pull" hypothesis of the ovarian cycle discussed in the text.

makes the follicles grow, ripens the follicles and eggs, and causes the production of estrogenic hormone. The rising tide of estrogenic hormone thus checks the production of pituitary hormone, which begins to fall off as estrus occurs. The estrogenic hormone is used up, and as it reaches a low ebb, the pituitary, now freed from the repressive action of the ovary, again begins to secrete its gonadotrophic hormone. Up goes the pituitary and then up goes the ovary again, thus getting another cycle under way. This scheme, however, cannot fully explain the cycle. As Lamport has shown, a push-pull action of the two hormones would naturally tend, not to effective cyclic fluctuations of estrogen, but to ever smaller changes approaching equilibrium. We must therefore postulate some other event which occurs from time to time to break the bal-

2 For a discussion of this theory, see Carl R. Moore and Dorothy Price, "Gonad hormone function." *American Journal of Anatomy*, vol. 50, pp. 13-72, 1932, and Harold Lamport, "Periodic changes in blood estrogen." *Endocrinology*, vol. 27, pp. 673-680, 1942.

ance of the two hormones. Under the push-pull hypothesis it is in fact easier to understand the long diestrous phase of cycles like those of animals that have an annual cycle, when (as we may suppose) the estrogenic and gonadotrophic hormones are balancing each other, than to explain what happens to set the see-saw swinging again, or to bring on cycles, in some animals, every few days or weeks. At present we can only make vague conjectures about the possible role of other hormones, e.g., progesterone or another pituitary hormone.

Events of the cycle. To resume the main theme of our discourse, during all these changes in the ovary the uterus is, of course, constantly under the influence of the ovarian hormones. Even when there is a long anestrous interval between one ovulation and the next, the ovary produces enough estrogen to protect the uterus from atrophy. As the follicles enlarge and ripen, there is a period of growth and development of the lining of the uterus (endometrium). When the corpus luteum is formed and begins to produce progesterone, the uterine lining is rapidly brought into the progestational condition, as described in Chapter V and illustrated in Plate XVII. About one week is required to complete these changes. The favorable environment thus prepared for the embryos (pages 107-111) is maintained about one week longer, making two weeks in all between the beginning and the end of the active life of the corpus luteum. This is the progestational phase of the cycle. If the animal mated while in estrus, the embryos arrive in the uterus about the 4th day (later in some species) and begin to attach themselves sometime between the 7th and the 13th day, according to the species. It will be seen that the corpus luteum functions long enough to give time for implantation of the embryos. If this occurs, some sort of signal, probably via the pituitary gland, causes the corpus luteum to survive and maintain the uterus in a state favorable to early pregnancy.

Retrogression of the uterine changes. We are considering

here, however, a cycle in which the eggs are not fertilized. In such a case they are transported through the oviduct to the uterus, where about 8 or 9 days after they first left the ovary they go to pieces and disappear. The corpus luteum holds on until the 14th or 15th day, then degenerates and ceases to deliver progesterone to the blood stream. The endometrium is thus deprived of its hormonal support. The changes induced by progesterone disappear in the course of a few days. The blood flow through the uterus diminishes, the lining becomes thinner, the cells of its surface epithelium and glands diminish in number and height, and the glands resume the simpler form that characterizes the interval and follicular phase of the cycle. Generally speaking, the steps of this reversion are gradual; it is spread over several days, and gives no outward sign to let us know it is in progress.

In our diagram (Fig. 19) the whole sequence of changes in the lining of the uterus is illustrated by the middle portion, which is a conventionalized representation of the glands in their successive phases.

The cycle in menstruating animals. The cycle of the menstruating animals and the human species is fundamentally similar to that of other animals. Two important differences, however, exist. In the first place there is not a sharply defined phase of sexual receptivity like the estrus of other mammals. Although cyclic fluctuations of sex activity occur in some of the apes and monkeys, this is by no means as well defined as in most other animals, and in the human female sex desire is obviously much more influenced by all sorts of moods, social situations, domestic ups-and-downs, and the like, than by any tendency to cyclic alternation. Mating may occur on any day of the cycle. There is no outward sign, like the estrous behavior of lower mammals, to indicate the time of ripening of the ovarian follicle and its egg.

It is interesting to speculate about the effect of this suppression of estrous rhythm upon human life and the progress of the race. Certainly our customs would be very different from what they are if the sexual compulsions of women were like those of animals with strongly marked estrous periods. In these creatures the sex response, intense and irresistible in the female during estrus, is wholly absent at other times; in the human species it is moderated but diffused over a larger proportion of the time. In their various aspects and sublimations, from downright sex desire to affection and vague romantic yearnings, the impulses of sex color in some degree our entire adult lives, teach us to love nature and art, and call us to sacrifice and devotion. In this respect above all mankind differs from the beast.

Regardless, however, of this all-important difference of behavior, the cycle of the ovary proceeds in the human species as in the others (Fig. 22). The follicle ripens and ruptures, the egg passes to the uterus, the corpus luteum forms and takes up its endocrine function. The lining of the uterus undergoes a profound progestational change. The epithelial cells of its glands multiply so greatly that the glands have to become sinuous and pleated, in order to be accommodated in the available space. The glands fill up with fluid secretion and therefore become dilated. The result is a very characteristic appearance, when seen in sections of the uterus. This progestational or "premenstrual" state is well shown in Plate XXI, C.

Inspection of the diagram (Fig. 22) will show that the premenstrual phase is at its height during the second week after discharge of the egg from the ovary, just as in other mammals. If there is a mating, and the egg is fertilized and becomes an embryo, it will reach the uterus when the endometrium is fully under the influence of the corpus luteum and ready to take care of the new arrival.[3] This is clearly illus-

[3] We do not actually know the time of arrival of the human embryo in the uterus, nor the precise time of its implantation, since no normal

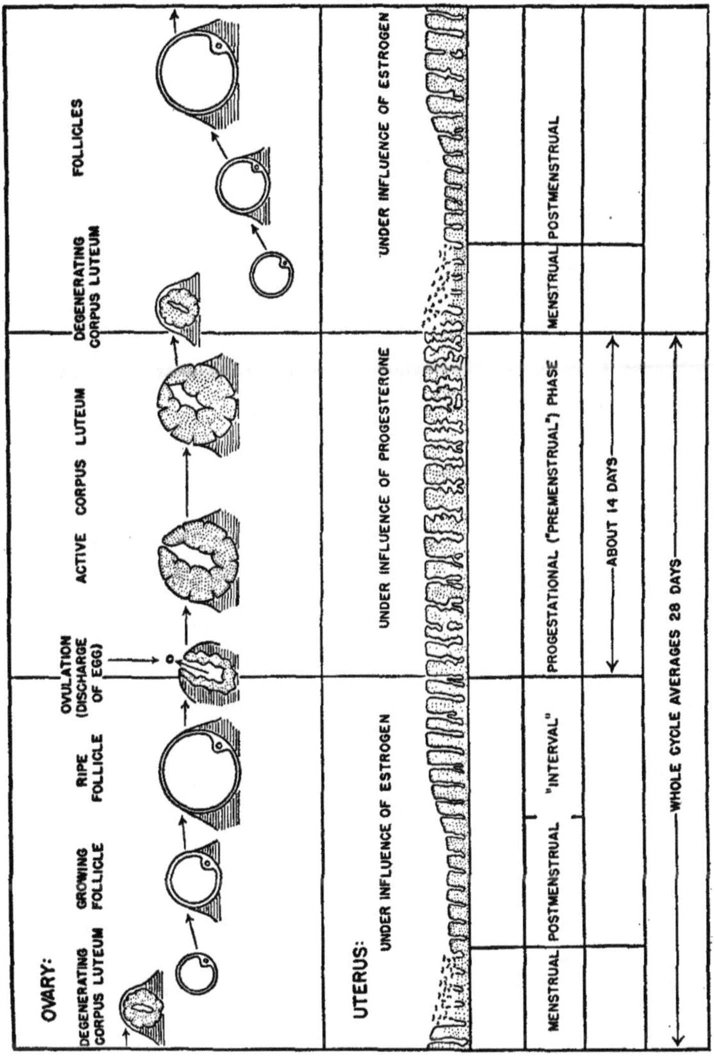

Fig. 22. Diagram illustrating the sequence of events in the menstrual cycle.

human embryos younger than about 7½ days have as yet been seen. Information from other animals, together with what is known of the human embryo at the 8th day, makes it highly probable that the human embryo becomes attached about the 7th day after ovulation.

trated in Plate XII, *C*, a photograph of one of the earliest known human embryos, obtained by Dr. Arthur T. Hertig of Boston, and preserved in Baltimore at the Department of Embryology of the Carnegie Institution of Washington. The uterus in which this 11-day embryo has attached itself is in the typical progestational phase, as shown by the form of the glands.

The menstrual breakdown. About the 14th or 15th day after ovulation, the corpus luteum begins to degenerate, as in other animals. The uterus, thus deprived of support by progesterone, undergoes a violent reaction. In its innermost layer the circulation of blood is disturbed, the surface epithelial cells, the glands and the connective tissue are damaged, and the tissues break down. Blood from small ruptured vessels fills the cavity of the uterus and trickles toward the vaginal canal. A section of the endometrium at this time shows a remarkable picture; the surface layer has sloughed away, and the stumps of the glands jut into the central mass of blood and cellular debris. In the course of a few days a process of repair sets in, the lost surface cells are replaced, the glands restored and the debris cleared up.

The various stages of the uterine cycle are shown in Plates XXI and XXII, which present a series of specimens from the Rhesus monkey.

The sequence of these events is summarized in the diagram, Fig. 22. From this it will be seen that ovulation takes place about the middle of the interval between the two menstrual periods. It is customary to count the days of the primate

PLATE XXI. Three stages of the cycle of the uterus of the Rhesus monkey. *A*, 15th day of cycle, just after ovulation; interval stage. *B*, 23d day of cycle. Effect of corpus luteum hormone appears in the glands; early premenstrual stage. *C*, 27th day of cycle. Menstruation due one day later. Full progestation (premenstrual) stage. All magnified 10 times. *A*, Corner collection (no. 2); *B*, courtesy of C. G. Hartman (H. 826); *C*, courtesy of G. W. Bartelmez (B. 123).

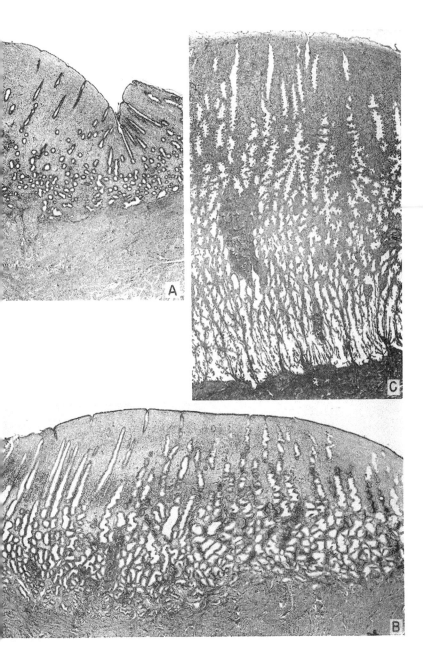

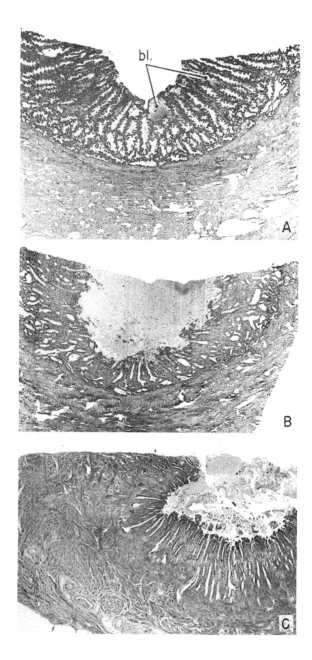

cycle from the first day of menstruation. Ovulation most commonly takes place about the 12th to the 16th day, although in individual cases it may be earlier or later than this. The corpus luteum is active for about 13 or 14 days, and therefore its degeneration brings on menstruation again 25 to 30 days after onset of the last period.

The "safe period." Let us digress again for a moment, to discuss, in passing, an interesting and important deduction that follows from the schedule of the human cycle, as shown in the diagram, Fig. 22. There is evidence from many species of animals that the eggs can be fertilized only while in the oviduct, during the first two or three days after their discharge from the ovary. We know also that the sperm cells cannot survive more than a few days in the female reproductive tract. It follows that the only part of the human cycle during which fertilization of the egg can occur is the few days following ovulation. Since, however, there is no way of ascertaining the date of ovulation and it may vary by several days, we shall for the sake of caution estimate the presumably fertile period as a few days longer each way, say from the 8th to the 20th day of the cycle, counting from the first day of the menstrual period. All the rest of the cycle, i.e. from about the 20th day to the 8th of the next cycle, will be a period of sterility, during which mating will not result in pregnancy. This is the theoretical basis of the so-called "safe period" method of birth control. If all women had regular cycles and things never happened out of turn, it would no doubt be a fully effective method, but irregularity

PLATE XXII. The uterus of the Rhesus monkey during menstruation. *A*, first day of flow; at *bl.*, small collections of blood in the lining of the uterus. *B*, third day; note loss of surface tissues of lining and disappearance of progestational pattern of glands. *C*, anovulatory menstruation, first day. Note, in comparison with *A*, that there is no progestational change of the glands. Magnified 10 times. *A*, *B*, Corner collection (nos. 39, 22); *C*, courtesy of G. W. Bartelmez (B. 128).

of cycles and unpredictable variations make it much less than certain.[4]

Role of the blood vessels of the uterus in menstruation. Within the past few years a good deal has been learned about what actually happens to produce the menstrual breakdown. Much of this advance we owe to G. W. Bartelmez of the University of Chicago, and to his former associate, J. E. Markee, now of Duke University. To make it clear we must first understand the arterial blood circulation of the lining of the uterus. The endometrium is fed by arteries which come up into it from the underlying muscle (Fig. 23). These have branches of two kinds. Those of one kind are very peculiar, for they are wound into coils, making their extremely tortuous way toward the surface, where they break

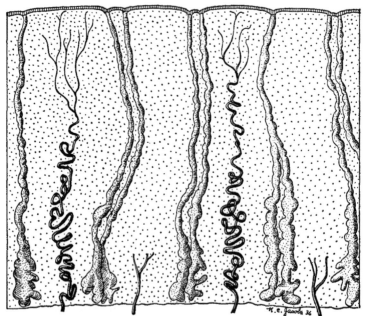

Fig. 23. Diagram of the arteries of the uterus, from the description of Daron. Enlarged about 20 times.

[4] Carl G. Hartman, *Time of Ovulation in Women.* Baltimore, 1936.

up into tiny capillary vessels (not shown in the diagram) that supply the inner one-third of the endometrium. The other kind of branching is that of the straight arteries, which run a short course directly to supply the basal two-thirds of the endometrium.

By studying under his microscope a series of uteri of women and monkeys, collected at successive stages of the cycle, Bartelmez showed that the fundamental step in the menstrual breakdown is a shut-off of the coiled arteries. With such material, however, it is possible to see only interrupted stages of the process; the sequence cannot be seen in full. Markee has therefore made use of a remarkably clever means of watching menstruation in progress.[5] Since we cannot see into the uterus, he undertook to put that organ (or rather, small pieces of its lining) into a situation where it can be watched. He grafted bits of endometrium into the anterior chamber of the same animal's eye, thus applying a method already used by a few investigators for other purposes. The small grafts are placed just behind the clear cornea, and get their blood supply through vessels which grow into them from the iris. The operation of grafting, which is done under complete anesthesia, is relatively simple, though, of course, it requires deft hands. The animal suffers no discomfort from the graft and is inconvenienced only by the fact that while under observation she has to sit in a tight wooden box, something like a pillory (but more comfortable), while the investigator studies her eye through a microscope. He is, by the way, at least as uncomfortable as the monkey, because the task of watching the winking, roving eye of the animal, changing the focus and moving the microscope and light whenever necessary, is enough to exhaust the patience even of a scientist.

[5] J. E. Markee, "Menstruation in intraocular endometrial transplants in the Rhesus monkey." *Carnegie Institution of Washington, Publication No. 518 (Contributions to Embryology,* vol. 28), pp. 219-308, 1940.

The grafts survive and grow. They respond to estrogenic hormone, injected under the skin, by swelling and growing just as if they were still part of the uterus. If the ovaries are removed, the grafts undergo castrate atrophy. Most astonishing of all, when menstruation occurs in the uterus, it occurs at the same time in the eye-graft, runs the same course, and ceases at the same time. The menstrual hemorrhage which occurs in the eye, stains and clouds the aqueous humor for a few days but soon clears away.

Markee was able to watch the process through the microscope, using low to moderate magnification, from 12 to 150 times. What he saw has helped us greatly to understand the nature of the menstrual breakdown, although (as we shall see) there is much still to be learned. Markee tells us that the first sign of impending menstruation in the eye-graft is blanching of the tissues due to shutting off of the blood flow by contraction of the coiled arteries. This does not happen in all the arteries of the graft at one time, but in individual arteries, so that blanched patches appear here and there in the graft, until all the tissues ultimately experience the blanching. After a few hours this phase wears off. Through the relaxed arteries the blood flows again with renewed force, but the tissues of the endometrium and especially the capillary blood vessels have suffered from the lack of blood supply. Here and there the small vessels give way and burst, causing tiny spurts of blood into the tissues. The little pools of blood thus produced coalesce and drain into the anterior chamber of the eye. In the uterus itself, similar hemorrhages are of course discharged into the cavity of the uterus. After a few days this strange series of events is over, and the damage is promptly repaired.

With Markee's direct observations to guide us, the study of prepared specimens of the uterus is much clearer. Observations by the two methods agree perfectly, but without observations of the eye-grafts we should probably not have learned the importance of the periodic shutting off of the spiral arteries.

Coiled arteries of the type thus shown to be fundamentally involved in menstruation in the monkey are also present in the human uterus, but have never been found in non-menstruating animals. Menstruation, then, is primarily an affair of the coiled arteries, which control the blood supply of the inner layer of the endometrium and by their closure cause breakdown, tissue damage, and hemorrhage (Appendix II, note 12).

In view of the violent disruption that characterizes the retrogressive phase of the cycle in women and in the other menstruating primates, it is a matter of great theoretical interest to know whether this stage in the non-menstruating animals is actually as free from tissue breakdown as I have rather summarily indicated. In other words, is menstruation a totally peculiar affair, sharply different from what goes on in mammals generally, or is it merely an exaggeration of a degenerative process that is present but not extensive in lower animals? This question is being investigated, but the answer cannot be given now. We need to know most of all what goes on in the uterus at the end of the corpus luteum phase in the New World monkeys (the capuchins, spider monkeys, and howler monkeys), which in spite of their close evolutionary relationship to the other primates do not menstruate externally. Here, if anywhere, we may expect to find transitional conditions that may help explain the wherefore of menstruation. The evidence is not yet in, but I may say that there are hints, apparent to the expert microscopist, that even in the rabbit and other non-menstruating mammals the retrogressive phase has an element of acute damage in it. These signs are, however, slight indeed and the statement holds true that in almost all mammals, when the corpus luteum has done its work, and the uterus is released from its phase of progestational proliferation, it

{ 153 }

settles gently and inconspicuously back to the state it was in before the follicles matured.

Theories about the menstrual cycle. There is no need to discuss outmoded theories of the cycle here, except to exclude one or two ancient fallacies that still crop up occasionally. For example, some people still consider that menstruation is equivalent to estrus. This is a common notion among farmers. Because menstruation is the most prominent event in the human cycle, and estrus the most conspicuous phenomenon in the cycle of the barnyard animals, they are wrongly considered to be fundamentally alike. It would follow from this that in humans the egg is shed from the ovary at the time of menstruation, a notion which is absolutely incorrect, as will be seen from our previous discussion. Menstruation is the last stage, not the first, of the corpus luteum phase of the cycle.

Another false view, which prevailed widely among European gynecologists from 1880 to 1910, asserted that there is no chronological relation whatever between ovulation and menstruation. The egg may be shed at any stage of the cycle. This conclusion was drawn by surgeons who knew very little about other species, and who moreover usually saw at the operating table not normal pelvic organs, but those of patients with gynecological ailments, often subject to disturbances of the cycle.

When it began to be understood clearly that the ovary is an organ of internal secretion, a group of first-class German gynecologists, including especially Robert Schroeder, Robert Meyer, and Ludwig Fraenkel (the latter two now in exile) developed a theory which had been vaguely outlined a generation earlier, that the corpus luteum is in some way associated with the menstrual cycle. Gradually their views, clarified by intensive observation of human material, arranged themselves into a theory of the cycle which was very plausible and which has turned out to be partly correct.

This states that menstruation is simply the downfall of the premenstrual (progestational) endometrium, and that it is caused by the degeneration of the corpus luteum. It will be seen that this theory fits all that has been said about the primate cycle thus far, and that it is compatible with our diagram, Fig. 22. According to this theory, the endometrium cannot menstruate unless it is first built up to the "premenstrual" state. Professor Meyer put this into an aphorism which was much quoted by the gynecologists "Ohne Ovulation keine Menstruation"—without ovulation there can be no menstruation.

This is a beautiful, clear hypothesis, and it is half true. It is also, unfortunately, half false. The fallacy is subtle but fundamental, and leads us headlong into a mass of unsolved problems.

Anovulatory cycles. The failure of the German theory of the cycle is a matter of especial interest to me, for it was my lot to obtain (to my great perplexity) the first undeniable evidence against it. The story is best told as it happened. In 1921, after several years of work on the cycle of the domestic pig, I felt prepared to begin a study of a menstruating animal and for this purpose I chose the Rhesus monkey. Practically nothing was known on the subject. There had been two investigations. Walter Heape, a distinguished English biologist, had gone to India more than twenty years before to study reproduction in Rhesus monkeys and langurs, but illness had forced him to return to Cambridge, where he followed and described the cycles of a few animals he had taken home with him. M. A. Van Herwerden had studied material of a wholly different kind. Hubrecht, the great embryologist of Utrecht, had collected a great many reproductive tracts (uteri with ovaries) from several species of monkeys. These had been obtained largely by Dutch colonial officers in the East Indies. Miss·Van Herwerden examined these specimens, which were unaccompanied by life histories or records of menstrual cycles, because the animals had been shot in the jungle by hunters. As regards the relation of menstruation to ovulation in the cycle of the monkey, the results of Heape and Van Herwerden were obscure and puzzling. Heape in his few cases observed no clear relation. Van Herwerden actually found that in some of the Hubrecht specimens the uterus was menstruating but there was no corpus luteum at all in the ovaries. In other menstruating animals a corpus luteum was present. This variability could perhaps be reconciled with the older theories of the human cycle, but not with the Meyer-Schroeder-Fraenkel theory. The absence of life histories, however, cast uncertainty upon the significance of Van Herwerden's observations. A hunter's specimen lets us see only one instant in the life of the animal; who could tell the significance of these puzzling cases so completely removed from the context of life?

Meanwhile the German interpretation seemed plausible indeed. It could be matched without difficulty to all the recently gained knowledge of the cycles of mammals. Stockard and Papanicolaou's studies of the guinea pig (1917), those of Long and Evans on the rat (1921), which I had been privileged to watch for four years, and my own on the domestic pig had all emphasized the occurrence of regular cycles of ovulation followed by the progestational phase of the uterus. I supposed that application of the same methods to a menstruating mammal, namely the Rhesus monkey, would reveal a strictly parallel sequence, with menstruation as its last stage. If I kept my animals in good condition, observed their cycles with perfect vigilance, and autopsied them at carefully chosen stages in their cycles, I should obtain a series confirming the German theory. I thought that 25 monkeys and three years' work would suffice to establish the normal cycle, after which we could go on to all sorts of experimental studies in confidence that we could elucidate the normal physiology and the disorders of the human menstrual cycle.

Imagine my confusion when the very first monkey we killed disagreed completely with all we had expected. Rhesus monkey No. 1 was in my colony more than a year. She had 12 menstrual cycles in 12 months, the last 5 of which were respectively of 27, 29, 25, 24, 27 days, averaging 26.4 days. In the hope of recovering a young corpus luteum and of finding an egg in the oviduct or uterus, she was killed 17 days after the onset of the last previous menstrual period and 9 days before the expected onset of the next. To our astonishment, neither ovary contained any sign of recent or impending ovulation. There was no large follicle, no recent corpus luteum, nor any older corpus luteum from the last two or three cycles. In short, this animal was undergoing cycles of menstruation without ovulation and therefore without corpora lutea.

Monkey No. 2, on the other hand, fulfilled our original expectations. She, too, had a series of regular cycles. She was killed 14 days after the onset of the last period and 12 days before the onset of the next expected period. The left ovary contained a recently ruptured follicle and the egg was in the oviduct. This, by the way, was the first egg of any primate ever recovered from the oviduct. The case fits the diagram perfectly.

To make a long story as short as possible, it turned out that Rhesus monkeys do not ovulate in every menstrual cycle.[6] When they do ovulate, the corpus luteum of course is formed and causes progestational (premenstrual) changes in the uterus. When the corpus luteum degenerates, typical menstruation occurs, by breakdown of the premenstrual endometrium. When the animal does not ovulate, then natu-

[6] George W. Corner, "Ovulation and menstruation in Macacus rhesus." *Carnegie Institution of Washington, Publication No. 332* (*Contributions to Embryology,* vol. 15), pp. 75-101, 1923.

rally there is no corpus luteum and therefore no premenstrual change in the uterus. Menstruation occurs anyway, and the breakdown takes place in an endometrium which is still in the unaltered state. The corpus luteum is necessary for the premenstrual state, but not necessary for the breakdown.

This analysis of the situation has been confirmed by Markee through watching menstruation in endometrium grafted in the eye. Markee tells us that when he applies the microscope to the grafts he sees only one difference between ovulatory and anovulatory menstruation, namely the occurrence of the progestational phase in the former and its absence in the latter. The shutting off of the blood supply, the subsequent reflux of blood through the coiled arteries, the rupture of the small vessels and the hemorrhage are the same in both instances. With all this evidence it can hardly be doubted that anovulatory bleeding is also menstruation.

It is not possible to distinguish between ovulatory and anovulatory menstruation by ordinary observation of the living animals. The cycles are of similar length, the bleeding is similar in appearance and duration. Recent studies by Inés de Allende and Ephraim Shorr suggest that it may be possible in the future to detect anovulatory cycles by studying the vaginal cells.

My description of menstrual cycles without ovulation was at first rather generally mistrusted, but it has been confirmed by everyone who has studied the Rhesus monkey.[7] We know that anovulatory cycles are likely to occur in young animals in the first months after the establishment of menstruation, and in fully mature females in the early fall and late spring, that is to say at the beginning and end of the active breeding season of the winter months. Rhesus monkeys do not menstruate regularly in summer. The anovulatory cycles tend

[7] Carl G. Hartman, "Studies in the reproduction of the monkey, Macacus (Pithecus) rhesus." *Carnegie Institution of Washington, Publication No. 433 (Contributions to Embryology,* vol. 23), pp. 1-161, 1932.

to occur, therefore, when the reproductive tract is preparing for its highest activity or receding from it. My papers describing the monkey cycle set off an active debate among the gynecologists as to whether anovulatory cycles occur in women. After much discussion and a great deal of careful observation, it is generally agreed that anovulatory cycles do occur, though with much less frequency than in monkeys. They seem to be most frequent in young girls and in women approaching the menopause.

There is no place for menstruation without ovulation, in the theoretical scheme which I have called the German theory. Therefore the savants who had formulated that theory simply declared that anovulatory menstruation is not menstruation at all. The rest of us, however, have gone on trying to find an explanation that fits all the facts. In this search the new knowledge of the ovarian hormones has begun to help us.

THE HORMONES AND MENSTRUATION

Experimental uterine bleeding. A simple experiment, made in 1927 by Edgar Allen, opened up the whole problem of the relation of the ovarian hormones to menstruation. Allen found that removal of both ovaries from a mature Rhesus monkey will usually cause within a few days a single period of menstruation-like bleeding. Why the medical profession had failed to discover this fact from human surgical patients is difficult to understand. It has long been known that removal of the ovaries abolishes the menstrual cycles, but the doctors had missed observing the fact that one period of hemorrhage often follows the operation before the cycles cease permanently. They seldom remove the ovaries except in the presence of disease, when the cycles are already altered, or for tumors which themselves produce bleeding, or as part of a larger operative procedure which may cause surgical hemorrhage from the uterus. Thus bleeding due purely to removal of the ovaries escaped notice until Edgar Allen discovered it in monkeys.

As a matter of fact, an experiment like Allen's had once been done on humans, on a large scale, and with the best intentions in the world. Robert Battey, a surgeon of Augusta, Georgia, in 1872 conceived the idea that neuroses and insanity in women are often concerned with the ovaries and may be treated by removal of these organs. He was probably led into the notion by observation of cyclic mental disturbance, paralleling the menses, insanity following child-bearing, and other conditions in which sex and the reproductive functions were of course concerned, though in a far more complex way than he could have imagined. Battey's radical proposal to remove the normal ovaries was put forward just at the time when the surgeons had gained command of the operation of ovariotomy (as they often ungrammatically called it). Antiseptic surgery, Lister's gift to the world, was now in general use, and the great American ovariotomists Ephraim McDowell, the Atlees, and their followers in Britain and Europe had worked out the operative technique. The operation was therefore relatively safe, and no doubt the patient's mental condition was often improved or at least subdued by the surgical intervention, with its anesthesia and opiates, by the rest in bed, and the nursing and general attention. At any rate "Battey's operation" was taken up widely by a profession thoroughly baffled by mental disease. Thousands of women were subjected to this drastic operation, not only in the United States, but in England, Germany and the rest of Europe, until in good time it became obvious that the psychiatric results did not justify it and that insanity with cyclic or sexual symptoms cannot be pinned directly to the ovaries. The late Dr. Edward Mulligan of Rochester, New York, told me of an incident in the last years of Battey's operation. Dr. Mulligan when a young surgeon studied for a time, about 1883, at Bellevue Hospital in New York City with the pathologist William H. Welch, himself a young man on his way to the Johns Hopkins and the leadership of the American medical profession. One morning Welch showed his pupils a tray containing a number of normal ovaries, removed that morning in the operating rooms, and took the occasion to denounce the practice of Battey's operation in words so vigorous that Dr. Mulligan still remembered them more than forty years afterward.

The point of all this is that removal of the normal human ovaries was very often followed within a few days by a period of bleeding from the uterus lasting several days. This appears in many of the case reports in medical journals from 1872 to 1885. The doctors did not always report all the postoperative details, but when they did they generally noted the hemorrhage, but never with comprehension. Thus an important observation was missed because the observers' minds were unprepared.

We must digress for a moment to mention that under the strict corpus luteum hypothesis of menstruation (which I have for brevity called the German theory) removal of the corpus luteum may be expected to bring on menstruation. This had been perceived and demonstrated at the operating table before 1927, when Allen announced, on the basis of his experiments on monkeys, the broader fact that removal of both ovaries, with or without a corpus luteum, has the same effect. It may help keep things clear, if we point out something the reader has probably thought out already, namely that removal of a corpus luteum produces bleeding from a premenstrual endometrium, whereas removal of the ovaries without a corpus luteum produces bleeding from an unaltered endometrium, as in anovulatory menstruation.

Estrin-deprivation bleeding. Edgar Allen reasoned that the effects of removal of the ovaries of his monkeys were really due to removal of the estrogenic hormone, which had recently been discovered, thanks so largely to his own investigations. He therefore took a castrated female monkey and gave her a course of injections of estrogenic hormone. When he discontinued the treatment, bleeding ensued. In other experiments he removed the ovaries and immediately began daily doses of estrogenic hormone. As long as the hormone was given, there was no bleeding; that is to say, the hormone was able to substitute for the ovary. When it was discontinued, the bleeding occurred.

The following diagram represents graphically the experiments just described.

Removal of ovaries; estrin deprivation:

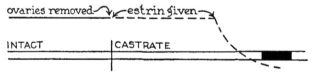

Fig. 24. Illustrating the experiments of Edgar Allen, 1927, 1928. In this and the following 3 graphs the black bars indicate uterine bleeding. These diagrams are from an article by the author in the *American Journal of Obstetrics and Gynecology*, by courtesy of the C. V. Mosby Company.

On this basis Allen formulated the estrin-deprivation hypothesis of menstruation, which suggests that natural menstruation, like the experimental bleeding, is due to a cyclic reduction of the amount of the estrogenic hormone available in the body.

Subsequent experiments done with carefully graded doses of the hormone, including especially those of Zuckerman, of Oxford, have shown that not only total deprivation, but also mere lowering of estrogen dosage below a certain level will produce the bleeding. The word "deprivation," as used in this connection, is therefore to be taken in a relative sense.

Estrin-deprivation hypothesis:

Test of the hypothesis:

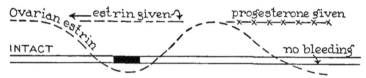

Fig. 25. In the lower figure the natural level of estrogen is shown fluctuating cyclically, not as a proved fact, but to show how the actual results of injecting the ovarian hormone bear on the estrin-deprivation theory.

The correctness of the estrin-deprivation hypothesis can be tested in a very simple way. We need only choose a monkey that is menstruating regularly, and keep up her estrogen level by injecting the hormone, for say ten days before the expected menstrual period. This should prevent menstruation. I tried this in a sufficient number of monkeys, giving them doses of estrogen I thought similar to their own natural supply from their ovaries. Later in the experiments these doses were increased several fold. The hormone did not stop the next menstrual period. If the treatment was continued into later cycles, menstruation was often delayed, perhaps because of roundabout action through the pituitary gland. Zondek has found that in women very large doses of estrogenic hormone disturb the menstrual cycle, owing to inhibition of the gonadotrophic mechanism of the pituitary. These upsets produced with long-continued or very large doses are however not altogether pertinent. The theory calls for relatively small fluctuations, within the body's normal range of hormone production. On this hypothesis, however, the very first period should be prevented, and this did not occur. If my dosage of hormone was really physiological, as we say (that is, something like the amount the animal herself would produce) then the estrin-deprivation hypothesis in its simple original form is not adequate to explain the observed facts.

Progesterone and menstruation. On the other hand, administration of the corpus luteum hormone even in small doses prevents menstruation in experimental animals, abolishing the very first menstrual period after injections are begun (provided they are started a few days before the expected onset). Both Hisaw and I have found that experimental estrin-deprivation bleeding, produced by removal of the ovaries or by discontinuance of a course of treatment with estrogenic hormone, is prevented by small doses of progesterone. Smith, Engle and Shelesnyak at Columbia University arranged an exceedingly vigorous condition of estrin deprivation. Their monkeys were first given a course of gonadotrophic hormone from the pituitary gland (see p. 141) to stimulate the output of estrogenic hormone from the ovaries. They were also given generous doses of estrogens for good measure. These hormones were suddenly discontinued and at the same time the ovaries were removed. In the face of all these reasons for deprivation bleeding, modest doses of progestin (crude progesterone) completely prevented hemorrhage.

On the other hand, progesterone deprivation, like estrin deprivation, invariably causes menstruation-like bleeding. This was very clearly apparent in a series of my experiments in which progesterone was given to normally menstruating monkeys. During the injections, menstruation ceased. When

Progesterone prevents estrin-deprivation bleeding:

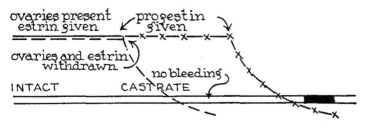

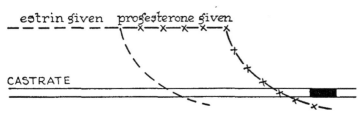

Fɪɢ. 26. The upper figure illustrates the experiments of Smith and Engle, 1932, and Engle, Smith and Shelesnyak, 1935. The lower figure represents the results of Hisaw, 1935, and Corner, 1938. Estrin-deprivation bleeding is postponed by progesterone; discontinuance of progesterone is then followed by bleeding.

the hormone was purposely stopped at a time which would have been midway between two periods (had the latter been occurring on their original schedule) progesterone-deprivation bleeding then occurred within a few days. The monkey next menstruated spontaneously about 4 weeks after the experimental period. This tells us that the monkey's time-piece mechanism accepts progesterone-deprivation bleeding as if it were actual menstruation, and takes a fresh start from the induced period.

The diagram (Fig. 26) illustrates these facts.

At this point Dr. Markee may be called again as a witness. He tells us that when bleeding is produced in one of his eye-grafts, by withdrawal of either the estrogenic hormone or progesterone, he observes the same sequence of blanching and hemorrhage that occurs in spontaneous menstruation. If the monkey is given progesterone, the graft will bleed from a "premenstrual" state; if given estrogenic hormone, it bleeds from the interval state.

With this information in hand it is possible to plan an experiment in imitation of the normal ovulatory cycle. This is represented in the upper half of the following diagram, Fig. 27. The underlying idea occurred to workers in the Oxford, Harvard, and Rochester (N.Y.) laboratories at

Progesterone during a course of estrin:

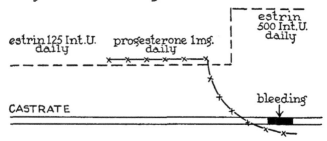

Current hypothesis of ovulatory menstruation:

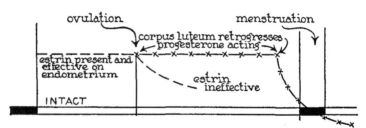

Fig. 27. The upper figure illustrates the production of bleeding after discontinuance of a course of progesterone, in spite of continued and more intensive estrogen treatment. The dosage shown is that of the author's experiments (Corner, 1937, 1938); similar results were obtained by Zuckerman, 1937, and by Hisaw and Greep, 1938.

practically the same time, and gave consistent results when tried. The dosage cited here is that of my own version of the experiment. A castrate female monkey is given a daily dose of estrogenic hormone, 125 international units, sufficient to build up the endometrium to normal thickness and structure. After 10 days, a daily dose of progesterone is added (just as would have happened had the animal developed a corpus luteum of her own). Ten days later, at the 20th day of the experiment, the progesterone is discontinued, but the daily injection of estrogenic hormone is continued. In spite of the estrogen, we find that bleeding invariably occurs in a few days. Indeed, the dose of estrogen may be greatly increased, say to 500 international units, beginning on the day on which the progesterone is discontinued; but menstruation-like bleeding still occurs. Seven hundred units or more may be necessary to prevent it, although such doses as 500 or 700 international units are of course much more than necessary to maintain the uterus when not working against progesterone deprivation.

These facts enable us to construct a relatively simple hypothesis of the menstrual cycle which is really a modified form of the estrin-deprivation hypothesis (Fig. 27, lower part). We start by assuming that progesterone in some way or other has the property of suppressing the menstruation-preventing power of estrogen, while itself holding off menstruation. In the normal cycle the animal does not bleed in the first half of the cycle (follicular phase), because the ovaries are furnishing estrogen. She will not bleed during the second half of the cycle (corpus luteum phase) because the corpus luteum is furnishing progesterone. By our assumption, however, the corpus luteum is suppressing the protective effect of the estrogen; therefore when the corpus luteum undergoes retrogression, the animal finds itself deprived of the action of both estrogen and progesterone, and the endometrium breaks down. Ovulatory menstruation is thus a special case of estrin-deprivation bleeding.

This explanation of the normal cycle is, of course, simply a hypothesis which has been formulated to explain our observations. Whether things happen this way in the normal monkey or in woman remains to be proved. It is at least not contradicted by any of the known facts, and it has the merit of simplicity, because it calls for a cyclic variation in only one hormone, namely progesterone. Its one unproved assumption, that progesterone somehow cuts down the action of estrogen on the uterus, is supported by various other evidences of a sort of antagonism between the two hormones in some of their other activities. It has been suggested that progesterone has the property of speeding the elimination of estrogens from the body. If this proves correct it is all we need to complete our hypothesis.

This scheme does not explain anovulatory menstruation, for in cycles without ovulation there is, of course, no coming and going of the corpus luteum. Anovulatory menstruation is therefore probably due to estrin-deprivation alone. We can imitate it perfectly in castrate animals by simply giving a course of estrogen injections interrupted or sharply reduced at suitable intervals. What is actually happening in the female organism remains to be worked out. We are not yet sure that there is an actual up-and-down of estrogen in the body sufficient to produce deprivation bleeding. The daily assay of estrogens in the blood is very expensive and at present not reliable enough for our purpose. Zuckerman has shown that when a castrate monkey is kept on a relatively small but constant daily dose of estrogen, there is a tendency to occasional uterine bleeding which may become fairly regular. We may conjecture, on this basis, that perhaps some sort of give-and-take relation exists between estrogen and some other hormone, just as between estrogen and progesterone when the corpus luteum is present, so that even without the corpus luteum a periodic state of estrin deprivation occurs. Can it even be that the adrenal gland produces something that can suppress the estrogens (we know that a number of steroidal compounds resembling progesterone are extractable from that gland)? The play of hormones in anovulatory menstruation is anybody's guess, and those of us who have worked on it can assure our colleagues, on the basis of much vain conjecture and many futile experiments of our own, that the problem is not an easy one. Some little fact is lurking just beyond our grasp.

Since the first draft of this chapter was written, Hisaw has reported from the Harvard zoological laboratory some experiments which show that very small doses of progesterone given and then discontinued (1 milligram a day, for one to five days) will set off menstruation-like bleeding in castrate monkeys which are receiving large daily quantities of estrogenic hormone. He suggests therefore that anovulatory menstruation may be due to progesterone deprivation; even if there is no corpus luteum, he says, there may be a little progesterone produced in Graafian follicles (there is, in fact some collateral evidence for this latter part of the conjecture) and this may be enough to cause menstruation when such a secretion of progesterone ceases. It is a plausible conjecture, and one which calls for no new factor outside the ovary; but it will be difficult to prove.

The immediate cause of the menstrual process. From the foregoing sections it will be perfectly clear that the breakdown and hemorrhage of menstruation are consequent to the deprivation of estrogenic hormone or progesterone. It is also very probable, from the studies of Markee, that these effects are initiated by constriction of the peculiar coiled arteries of the endometrium, which produces damage to the tissues and ultimate degeneration. But how can it be that a temporary deprivation of one of these two particular hormones can shut off the arteries in one particular tissue?

This question is now the key problem in the theory of menstruation, and it remains unsolved. The attack upon it is in the stage of skirmishing, in which all the possible explanations are being put forward for discussion and trial; but up to the present no one of them has found support by experiment. The reader may, however, be interested in the mental processes of a group of puzzled investigators, and therefore I list the conjectures for what they are worth. To merit consideration at all, any explanation must fit the facts that (a) hormone withdrawal causes uterine bleeding; (b) this does not take place at once, but only after 3 to 8 days; (c) the bleeding, once hormone deprivation is well under way, cannot be postponed by renewed injections of estrogen or progesterone; (d) it takes place in grafted bits of endometrium (in the eye or elsewhere) which have no connection with the nervous system.

Since all of these conjectures involve the arteries, it may be helpful to recall the fact that the walls of an artery contain numerous cells of involuntary muscle, laid on in circular fashion around the inner tube (endothelium) that conducts the blood. When these muscle fibers contract, they squeeze down upon the blood stream like a man's fingers about a rubber bulb. It is thus that the blood pressure is raised by a dose of adrenin or by a strong emotional state, both of which cause the arterial muscle cells to contract. Such muscular cells exist in the coiled arteries of the uterus as in all other arteries (Appendix II, note 13).

Hypothesis 1. It may be that the coiled arteries are peculiarly and directly dependent upon the ovarian hormones, in some such way (for example) as the ovary is dependent upon the pituitary. This means that withdrawal of the ovarian hormones would let down the condition of the coiled arteries, causing them to contract. This hypothesis is the simplest, calling for no other hormones or special substances, but it is exceedingly difficult to try out, for the only way of proving that the ovarian hormones are the sole factors involved is to exclude all other possible factors, but when we cut off the estrogenic hormone, how can we know we are not thereby putting some other factor into action? If somebody could show us how to keep a coiled artery alive and working outside the body where we could deal with it alone, we could soon test this hypothesis, but the infant art of tissue culture has by no means reached the point of keeping an artery alive all by itself, and moreover these arteries are so tiny that they would have to be handled under the microscope—a truly difficult project!

Hypothesis 2. This admits the possibility, mentioned above, that withdrawal of estrogen permits something else to go into action. We know that even in the non-menstruating animals, withdrawal of the ovarian hormone causes a certain amount of deterioration of some of the cells of the surface epithelium and of the glands of the uterus. Under the microscope we see fragmentation of the nuclei and the accumulation of protoplasmic debris in the cell bodies. Is it possible that some chemical substance produced in the course of cellular breakdown (as histamine, for instance, is produced in burned tissues) diffuses through the endometrium to the arteries and causes them to contract? This hypothesis has interested me very much and I have made many experiments to test it, but always with negative results.

Hypothesis 3. Another conjecture, a variation upon the foregoing, is that the uterine coiled arteries are sensitive, when not protected by the ovarian hormones, to some substance that is normally present in the blood stream. We must suppose that withdrawal of the hormone allows this substance to act upon the arteries. One of the possible constrictor substances would be pituitrin, the secretion of the posterior lobe of the pituitary gland, a hormone which is highly potent in promoting contraction of smooth muscle. Carl G. Hartman tried this with negative results, and moreover P. E. Smith obtained bleeding by estrin deprivation in monkeys from which he had removed the whole pituitary gland. Adrenin has been thought of, but Edgar Allen succeeded in producing estrin-deprivation bleeding in monkeys from which the adrenal glands had been removed. This experiment is not quite conclusive, for monkeys may possibly have other sources of adrenin beside the adrenal gland. Up to the present, at least, this hypothesis has yielded no valuable clues.

Hypothesis 4. George Van S. Smith and O. W. Smith have suggested that the bleeding of menstruation is caused by conversion of estrogenic hormone into a non-estrogenic by-product which is toxic to the endometrium. This hypothesis could perhaps explain ovulatory cycles, but it cannot be fitted very well to simple estrin-deprivation bleeding; in any case it will be acceptable only when somebody comes forward with chemical derivatives of the estrogenic hormones that are especially toxic to the endometrium (Appendix II, note 14).

These are the most plausible current guesses about the immediate cause of uterine bleeding after hormone deprivation. None of them has been proved or even rendered likely, by experiment. It is indeed vexatious that we cannot clear up this important problem.

The modern concept of the cycle. By way of summary, let us now set forth a concise description of the primate cycle as revealed by recent research. What follows will, I think, be accepted by most of the American investigators, and also the British, although a die-hard English gynecologist a few years ago dubbed it "the American theory."

We begin by suggesting that there is a basic tendency to cyclical function of the ovaries, and that this is produced by a sort of reciprocal "push-pull" reaction between the pituitary and the ovarian hormones, as explained in full earlier in this chapter. In most cycles of fully mature human females, the pituitary gonadotrophic hormones cause the ripening of a Graafian follicle, the discharge of its egg, and the formation of a corpus luteum. This in turn sets up the progestational or "premenstrual" state of the endometrium. When the corpus luteum degenerates, menstruation ensues because of the withdrawal of progesterone. In some cycles, however, especially in young girls and in women approaching the menopause, a follicle does not ripen in the ovary, but after about the same interval as in an ovulatory cycle, namely 4 weeks, some process not yet understood leads to reduced action of estrogenic hormone, and bleeding ensues which we call anovulatory menstruation.

This view of the cycle requires, of course, much more investigation before we shall be able to understand the whole process; I have already pointed out the larger gaps in our knowledge. It certainly represents a great advance toward the truth; and what is indeed important, it gives us a far clearer and more hopeful viewpoint about the disorders of menstruation than do older concepts of the cycle. Since menstrual bleeding is caused by fluctuation in levels of the ovarian hormones, it follows that noncyclic, pathological bleeding, such as occurs in excessive and irregular periods may also be caused by abnormalities in amount proportion, or kind of these hormones. We must consider that there are not two sharply distinct kinds of functional bleeding, one being normal menstruation and the other abnormal hemorrhage. On the contrary, the modern hypothesis tells us that we must expect a series of types of hemorrhage ranging from normal menstruation through every grade of disturbance to the most severe disorder of the cycle. Gynecologists are already beginning to study and treat these distressing and difficult conditions in the light of this concept, and we may well hope these same hormones that control the normal cycle will help us to control its aberrations and at last to banish the specter of uterine hemorrhagic disease.

THE UNKNOWN SIGNIFICANCE OF MENSTRUATION

In all this discussion of the nature and the course of menstruation, we have had nothing to say about the significance and possible usefulness of the periodic breakdown and hemorrhage. The human mind has an intractable desire that natural phenomena shall be useful. We are not comfortable in the presence of useless or undirected activity. Menstruation in particular ought to have a practical reason for its occurrence, for otherwise it seems a totally wasteful, destructive, and vexatious business. Up to the present, however, no one has been able to demonstrate such a meaning. There are in fact only two guesses that are even worth talking about.

The late Walter Heape of Cambridge, England, one of the pioneers in the study of the reproductive cycle, proposed in 1900 that menstruation is the same thing as a period of bleeding that occurs in female dogs when they are going into heat, i.e. in the week preceding ovulation. A somewhat similar proestrous bleeding occurs in cows, particularly in heifers. Such bleeding is easily explained, for it is clearly due to engorgement of the blood vessels produced by a strong action of the estrogenic hormone. Under the microscope it does not resemble menstruation; the blood oozes from superficial blood vessels and there is little or no breakdown of the tissues. When Heape wrote, nothing was known of the time of ovulation in the primate cycle, nor of the premenstrual endometrium and its dependence upon the corpus luteum. In the primates, on his theory, ovulation would be expected to occur during menstruation, or immediately after the flow, just as in the bitch and cow ovulation occurs about the end of the proestrous bleeding. Since we know now that ovulation takes place a week to ten days after the cessation of menstruation, we can reconcile Heape's theory with the facts only by supposing that in the menstruating primates there is first proestrous bleeding, then a delay unknown in the other animals, and finally ovulation. This theory, and one or two ingenious variations upon the same theme by later English writers, F. H. A. Marshall and Zuckerman, all seem too complex to be probable, and moreover they suffer from a very serious objection. Unlike the menstruation-like bleeding, which can be produced in monkeys and women by withdrawing estrogenic hormone, the proestrous bleeding of dogs is produced by building up the level of estrogenic hormone, as was shown by R. K. Meyer and Seiichi Saiki in our Rochester laboratory in 1931.

A few years ago, Carl G. Hartman discovered that in Rhesus monkeys there is almost always a slight bleeding from the uterus about the time of ovulation. This does not show externally and is discernible only by applying the microscope to washings of the vagina. Sections of the uterus made at this time reveal that a few red blood cells are escaping from superficial capillary blood vessels of the endometrium, which are engorged by the action of the estrogenic hormone. In the laboratories we call this slight bleeding "Hartman's sign" and take it as evidence that there is a ripe follicle in the ovary. This is the actual equivalent of Heape's proestrous bleeding. It has since been found to occur in women, though probably not as regularly as in monkeys. These observations prove clearly that menstruation is something else than proestrous bleeding.

Hartman has proposed another explanation for menstruation. He points to the fact that in many mammals, at the time of implantation of the embryo, the uterine secretion contains red blood cells. He tells us also that in many viviparous animals lower than the mammals, for example certain salamanders and fish, in which the embryo depends upon the maternal tissues for nourishment, bleeding in one form or another usually occurs into the brood chamber. Hartman compares menstruation to bleeding of this kind and conjectures that it is simply a means of getting the vitally useful blood pigment, hemoglobin, into the region where the early embryo is to reside. We have to suppose that if in any given cycle an egg is fertilized, the premenstrual changes go to the very verge of menstruation, letting a little blood out of the vessels into the tissues to enrich the implantation site for the embryo. Since, however, a "missed period" is the first sign of beginning pregnancy, we must also suppose that attachment of the embryo then stops the process before external hemorrhage occurs, and even before there is any significant breakdown of the endometrium. If there is no embryo, the breakdown goes all the way. This is a very interesting conjecture, for it assigns a necessary and worthy function to the strange flux of menstruation. Careful study, however, of the uterine lining of the monkey just before and during implantation of the embryo, and of the very few human specimens during early implantation that are as yet available, does not support this hypothesis, for they do not show beginning hemorrhage in the endometrium. We know that many other mammals succeed in implanting their embryos without any such provision of free blood or hemoglobin in the endometrium; and we know also that sometimes in women and often in Rhesus monkeys menstruation occurs in anovulatory cycles, when there can be no embryos to profit by it. This hints that perhaps the process of menstruation evolved without reference to the embryo.

Menstruation, then, is still a paradox and a puzzle—a normal function that displays itself by destruction of tissues; a phenomenon seemingly useless and even retrogressive, that exists only in the higher animals; an unexplained turmoil in the otherwise serenely coordinated process of uterine function.

ENDOCRINE ARITHMETIC

"*I often say that when you can measure what you are speaking about and express it in numbers you know something about it; but when you cannot measure it, when you cannot express it in numbers, your knowledge is of a meagre and unsatisfactory kind: it may be the beginning of knowledge, but you have scarcely, in your thoughts, advanced to the stage of* science."—LORD KELVIN, *in Popular Lectures and Addresses,* lecture on *Electrical Units of Measurement,* 1883.

ENDOCRINE ARITHMETIC

HOW much of a given hormone is found in the body at one time? How much is produced in a day? How much is found in the gland at one time? What is the output of a single cell? These are questions to which we must have an answer if we want to understand the glands of internal secretion and use our knowledge for the benefit of mankind. Certainly every merchant must have this kind of information about his stock in trade, and the manufacturer about his materials and his product. The science of endocrinology is, however, a long way from any such basis for calculation. How can we measure the output of a factory (i.e. an endocrine gland) when we do not know exactly what raw materials it uses, how it makes the product, or what becomes of the product when it is used? In the case of most of these chemical factories we do not even know the capacity of the manufacturing plant. The insulin factory, for example, consists of many thousand bits of tissue, the pancreatic islets, irregular in size and shape, scattered through the pancreas. In the pituitary, although the gland is measurable as a whole, we do not know what particular cells are associated with the various hormones made in the gland, nor even indeed just how many hormones it produces. So it goes throughout this puzzling system of glands. We are dealing at present largely with unmeasurable organs and with incalculable processes. We are able only to appreciate some of the end results, not the fundamental steps. To measure and calculate what is going on within the glands and thus to understand the chemical reactions and strike the balance of input and outgo—that task lies ahead.

RATE OF SECRETION OF THE CORPUS LUTEUM HORMONE

It happens that of all the endocrine glands, the corpus luteum is the one with which we can go farthest at present toward calculating the answers to the questions with which this chapter opens. The amount of glandular tissue can be ascertained, for it occurs in discrete masses of more or less spherical form and is composed of cells which can be measured and even counted with a fair degree of accuracy. The chemical structure and molecular weight of the hormone are fully known, and the dose necessary to produce certain definite effects is known. With these data at hand, let us take pencil and paper and see how far we can get.

The following calculations are of course approximate, not precise. They represent a preliminary exploration in a brand new field of study. There are unavoidable flaws in our procedure. In our arithmetical operations, for instance, we shall have to combine figures obtained from observations on rabbits with others learned from swine, thus violating one of the primary-grade rules of arithmetic, picturesquely stated long ago by one of my teachers, "You can't add cows and horses." Some other uncertainties will appear as we go along. If our results come out anywhere between 50 per cent and 200 per cent of the true figures we shall be doing very well for a start. Physicians who have to decide on the dosage of ovarian hormones for their patients will be glad to have even that; and as for my lay readers, they may at least find this chapter amusing. At any rate, when I presented some of these calculations at the Cold Spring Harbor Symposium on Quantitative Biology in 1937, that earnest little assembly of research men surprised itself—and me—by being very much amused, partly perhaps because it really was funny to see a medical man so gaily slip the leash and wander down a strange pathway, and partly because of the incongruity between our simple experiments on rabbits and sows, and the final emergence into pure theory in terms of molecules by the billion. Toward the end of my discourse, however, the hearers settled down and discussed the calculation quite seriously. Our chairman, in fact, sat up all the next night figuring on a further stage of the investigation, and in the morning, weary but still enthusiastic, brought me several more pages of arithmetic. What follows here is the substance of that presentation, revised in consideration of some facts contributed in discussion by members of the symposium, and corrected also in the light of subsequent knowledge.

1. *How much progesterone is produced daily by a given amount of corpus luteum tissue?* We begin by considering the fundamental property of the corpus luteum, namely, to produce progestational proliferation of the lining of the uterus (Chapter V, page 107 and Plate XVII, *B*). This has been studied chiefly in the rabbit, in which species it shows up with especial clearness. In a rabbit there are, of course, several corpora lutea at a time. By cutting down their number by surgical operation on the ovaries under anesthesia, it is possible to ascertain the minimum number of corpora lutea necessary to produce full progestational proliferation as in figure *D* of Plate XVIII.

(a) A young Frenchman, Joublot, the first to try such an experiment (1927), found that two corpora lutea were sufficient, but one was insufficient. I tried it also, and found that one corpus was sufficient, but half a corpus insufficient. Lucien Brouha, then in Belgium (1934), found one or two corpora lutea necessary. In our calculations we shall take the average of these three results and start with the assumption that in the rabbit one corpus luteum produces just sufficient progesterone to cause typical effects upon the lining of the uterus.

(b) To produce the same effect with progesterone, giving one injection per day, requires about 0.2 milligrams daily. Dr. Pincus reports that by giving the hormone in two injections, the daily dose can be brought down to about 0.13 mg. Let us take this lower figure as the basis for subsequent calculations. It may help to visualize the quantities we are discussing if we remind ourselves again that an ordinary postage stamp weighs 60 milligrams.

From (a) and (b) we may conclude that *one corpus luteum produces about 0.13 mg. of progesterone daily.* Here we are making another assumption, namely that the progesterone we administer in oily solution is utilized by the rabbit as efficiently as she can utilize the progesterone she makes in her own ovaries. This is really not impossible, for only very small amounts of oil are used and they are slowly but completely absorbed. This point will be discussed again later when we attempt to calculate the rate of output of estrogenic hormone.

Our calculation can be checked roughly by considering another action of progesterone, namely the maintenance of pregnancy in the rabbit. This requires more of the hormone. Brouha found that to preserve pregnancy to the 8th day, the rabbit must have three or more corpora lutea (instead of merely the one corpus luteum required to produce typical changes in the uterus). If we want to maintain pregnancy with progesterone after removal of the ovaries, we must also increase the dose something like 3 times or more. Willard Allen found 0.5 to 1.0 mg. of progesterone daily to be the necessary dose. In this sort of experiment we find therefore that one corpus luteum provides the equivalent of 0.16 to 0.33 mg. of progesterone daily, a figure in crude agreement, at least, with that of 0.13 mg. arrived at previously.

The volume of tissue in a rabbit's corpus luteum is approximately 3.25 cubic millimeters. Dividing this into the daily output of 0.13 mg., we get the answer to question 1: *One milligram of rabbit's corpus luteum produces about 0.04 milligram of progesterone daily.*

2. *How much progesterone is made daily in the ovaries of one rabbit?* Although the number of eggs shed by a rabbit at one time, and hence the number of corpora lutea in one crop, varies from 1 to 18, the number occurring most frequently (modal number) is 8. Eight corpora lutea multiplied by 3.25 (the volume of one corpus luteum in cubic millimeters) and then by 0.04 (the output of progesterone, in milligrams, by one milligram of corpus luteum) gives the answer to question 2: *The modal daily output of progesterone by the ovaries of one rabbit is about 1.04 milligrams*, and the range is from 0.13 mg. when only one corpus luteum is present, to 2.3 mg. with the maximum number of corpora lutea, namely 18.

3. *How much progesterone is produced daily by the ovaries of a sow?* This is important to know because the sow is the source of most of the natural progesterone that has been extracted and the only animal in which we know the amount that is present in the ovaries at any one time. The volume of one corpus luteum of the sow is approximately 525 cu. mm., the equivalent of 160 rabbit corpora lutea. Assuming that the corpora lutea of the two species are equally efficient, volume for volume, and therefore that the amount of progesterone produced in each is directly proportional to the amount of glandular tissue (an assumption for which at present there is no evidence) then 1 corpus luteum of the sow would produce 21 mg. of progesterone daily.

In a sow possessing the modal number of corpora lutea, which is 10 in that species, *the total daily output of progesterone would be 210 mg. per day.* The range would be from 21 mg. when one corpus luteum is present to 525 mg. with the maximum recorded number of corpora lutea in the sow, namely 25. This result seems improbably high, but since we have no present means of improving it, let us use it tentatively in answering the next question.

4. *How does the daily output compare with the amount present in the ovaries at any one moment?* All the corpora

lutea in the ovaries of one sow, when the modal number is present, weigh about 5 grams. By direct extraction the yield of crude progestin is equivalent to 0.05 mg. of progesterone per gram of raw tissue. Therefore one sow having 10 corpora lutea has about 0.25 mg. progesterone in the ovaries at the moment of killing. If the total output per day, estimated in section 3 above as 210 mg., is anywhere near correct, this means that *the amount present in the ovaries at any one moment is less than 2 minutes' supply*. This is a very important fact, for it points to a very high rate of "turnover" of secretion in the gland. We must think of the corpus luteum as making the hormone quite rapidly, working up only a little at a time but putting it through very quickly.

5. *What is the daily output of the human corpus luteum?* The volume of secretory tissue in the human corpus luteum is very difficult to measure, since the gland has a folded wall about a large cavity filled with connective tissue, like the monkey's corpus luteum shown in Plate IX, *A*. Estimates which I have made by two rough methods give 450 to 500 cu. mm. as the volume of secretory tissue in one corpus luteum. This is 150 times the volume of the rabbit's corpus luteum; and assuming once more that the human corpus luteum produces progesterone at the same rate as the rabbit's, volume for volume, then *the daily output of the human corpus luteum may be estimated as about 20 milligrams per day*.

This calculated result seems rather high when compared with a few estimates obtained in other ways. For example, Kaufmann in 1935 reported producing a progestational endometrium in a human patient with 50 rabbit units in 15 days. These were presumably Clauberg units, of approximately 0.4 mg. each, making the dose about 1.3 mg. daily. Assuming that this is somewhere near the minimum effective dose, and allowing for a ratio of about 1:4, as in the rabbit (section 1 above) between the minimum daily dose for progestational proliferation and the amount necessary to carry out the real task of the corpus luteum, namely maintenance of pregnancy, then the human corpus luteum would be expected to produce about 5 mg. daily. Wiesbader, Smith and Engle of Columbia University Medical School (1936) found that a certain effect of the removal of the human corpus luteum (bleeding from the uterus) cannot be prevented by substituting 0.5 mg. daily of progesterone by injection, but can be prevented by 5 mg. This suggests that 5 mg. is a quantity sufficient to produce one of the known effects of the corpus luteum, though not necessarily the full effects.

Another way of getting at this figure is through the fact that in human females, used-up progesterone leaves the body through the kidneys as sodium pregnanediol glycuronidate, as explained in Chapter V, page 120. Venning and Browne, who discovered this fact, found that if they administered a given quantity of progesterone, they could recover about half of it in the urine as the excretion product. When they collected all the pregnanediol excreted by a patient in a whole menstrual cycle, they found that the total recoverable amount was normally about 60 mg. This would mean about 120 mg. of progesterone actually produced. Since the corpus luteum is probably actively functional during about 10 days of each cycle, we arrive at an estimate of 12 mg. of progesterone produced daily. Another research group, Pratt and Stover of the Henry Ford Hospital, Detroit, obtained considerably smaller values, for their patients yielded only 2 to 3 mg. of pregnanediol daily, which we may consider to represent at the most 6 mg. of progesterone. It is known, however, that the chemical recovery of pregnanediol and estimations of corpus luteum activity based upon this method are subject to numerous errors not fully understood. It is perhaps all we should ask for, that the various estimates and calculations we have made and cited fall within limits as close as 5 and 20 milligrams per day (Appendix II, note 15).

Physicians who have used progesterone for disturbances

of menstruation and for threatened abortion have seldom administered more than 2 mg. per day. If (as seems credible) beneficial results have been obtained with this and even smaller dosage, we must suppose that clinical benefit may require only the redressing of a slightly disordered balance. Our calculations make it clear, however, that larger doses, of the order of 5 mg. or more, will have to be given thorough trial before the medical possibilities of this hormone are fully understood.

6. *What is the progesterone output of a single cell?* Returning to the rabbit's corpus luteum, it is possible to calculate approximately the output of a single cell.

The averages of a number of measurements of the diameters of individual corpus luteum cells were 0.028 x 0.028 x 0.036 mm., giving for the cell a calculated volume of 0.000015 cu. mm. Dividing this into the volume of the whole corpus luteum we get 217,000, and making due allowance for space occupied by the blood vessels we arrive at an estimate of 180,000 endocrine cells in one corpus luteum of the rabbit.

Since 180,000 cells produce 0.13 mg. progesterone per day, *the daily output of one cell is about 0.0000007 mg.*

The figure at first sight seems very small, but when the number of molecules is considered the resultant expression looks like the astronomer's rather than the biologist's quantities. We get the number of molecules made by a single cell, by the following calculation. We ascertain the molecular weight of progesterone by adding the atomic weights of the elements it is made of, namely 21 atoms of carbon, 30 of hydrogen, and 2 of oxygen. The sum is 314. This is the relative weight in comparison with the atomic weight of oxygen taken as 16. Applying Avogadro's law we know that 6×10^{23} molecules[1] weigh 314 grams. Dividing the latter by the former

[1] In dealing with very large numbers and very small decimal fractions it is convenient to avoid writing dozens of ciphers by using exponents. Thus 100 is 10^2 and .01 is 10^{-2}. 500 is 5×10^2. The figure cited above, 6×10^{23}, when written in full is 6 followed by 23 zeros, or six hundred thousand billion billions.

figure, the actual weight of one molecule is 5.2×10^{-22} gram. Dividing this weight into the weight of the daily output of one cell, we find that *one cell produces about 1.3×10^{12} or 1,300,000,000,000 molecules*, i.e. more than a thousand billion molecules of secretion produced in one day by one cell.

For comparison, it may be noted that in one cubic centimeter (about 1/3 of a thimbleful) of air there are about 10^{19} molecules.

What has just been presented is to the best of my knowledge the first attempt to calculate the actual output of a single secretory cell in any organ. It is, of course, no more than a first approximation to the truth. Yet such conjectures as this, improved and extended beyond the present powers of science, are going to lead us some day to the innermost secrets of cellular life. Perhaps the reader begins to be confused by all this reckoning of enormous numbers of very small things. Perhaps, on the other hand, he has acquired an awesome sense of the complexity of the cells, each one of them an island universe, a frail microscopic enclosure within which arise whirling billions of molecules, themselves in turn complex frameworks of latticed atoms. Upon the correct behavior of these fantastic congeries of particles in the corpus luteum each one of us depended for life itself when we were embryos; so did all our mammalian ancestors and so will our descendants forever.

QUANTITATIVE ASPECTS OF THE RECEPTOR ORGAN

We can get a further glimpse of the workings of the corpus luteum by doing a little figuring about the receptor organ, that is to say the uterus, which receives the progesterone and is affected by it.

As we have seen (Chapter V) the corpus luteum acts upon the epithelial cells which cover the inner surface of the uterus and dip down to form the uterine glands. It is the growth and multiplication of these cells which constitute the progestational proliferation by which the early embryos are nourished and implanted.

From a count of the epithelial cells in sections of the uterus, I estimate that a rabbit's uterus contains (in the two horns) about 100,000,000 epithelial cells. This means that each of the 180,000 epithelial cells in a corpus luteum can affect about 500 epithelial cells. The corpus luteum cells are, however, about 25 to 50 times as large (in volume) as the epithelial cells, and therefore each of the former takes care of about 10 to 20 times its volume of the latter at the start of progestational proliferation. The disproportion is not quite so great if we calculate on the basis of the amount of corpus luteum tissue to maintain pregnancy, not merely to induce progestational proliferation. On this basis one corpus luteum cell controls at first about 3 to 5 times its volume of epithelial cells. By the time the effect is complete, the number of epithelial cells has increased a great deal, and thereafter the corpus luteum cell must maintain more of them. This means that small amounts of progesterone stir up a proportionately large effect in the uterus. That sort of trigger-like action ("You push the button, we do the rest") is characteristic of the internal secretions. It is necessary to suppose that some sort of chemical reaction between the hormone and something in the cells of the uterus starts a chain of secondary reactions which profoundly change the physiology of the uterine lining. We might have a better idea of what actually happens if we could know how much progesterone actually reaches each epithelial cell; but this cannot be calculated, because we do not know what proportion of the hormone is diverted to other receptors (the uterine muscle, the mammary gland cells, and possibly other tissues). If all of it went to the epithelial cells of the uterus, each such cell would receive daily about 1.3×10^{-9} milligrams. This is almost 10^{12} molecules. The actual share received must be considerably smaller.

THE RATE OF SECRETION OF ESTROGENIC HORMONE

It is not possible at present to estimate the rate of secretion of estrogenic hormone with anything like the fullness and probability with which such an estimate could be made regarding the corpus luteum. As pointed out in Chapter III, the cells which produce the estrogenic hormone are probably scattered throughout the ovary in the walls of small and large follicles. The number and the total volume of these cells is obviously not subject to computation, and therefore the best we can do is to estimate the activity of the ovaries as a whole. I shall give here a summary of such an estimation which was published recently in more technical form elsewhere. As in the case of the corpus luteum, we get our answer by finding out how much of the hormone we need to administer in order to restore one or another of the functional effects of the ovaries after they have been removed.

How much estrogen is required daily by the monkey? As described in Chapter VI, removal of the ovaries of the Rhesus monkey causes menstruation-like bleeding from the uterus due to removal of the source of estrogenic hormone. If we take out the ovaries and then administer estrogenic hormone, we can try various doses and discover how much we have to give to prevent bleeding. This amount will represent the replacement of the hormone formerly produced by the ovaries when they were present. In my notebooks there are records of twelve experiments of this kind which are suitable for our present consideration. In each of them the animals received 125 international units of estrone daily, beginning the day the ovaries were removed. Of these, 4 bled from the uterus beginning on various days from the 3d to the 14th after the operation, in spite of the treatment with estrone. Seven did not bleed at all during the 15 days of the experiment. One, which received a larger dose of estrone, namely 500 international units, showed a few red blood cells from time to time, beginning on the 3d day, but no external bleeding. The experiment shows that 125 international units of estrone will substitute in many cases, but not in all, for the natural product of the animal's own ovaries, as tested by the prevention of postcastration bleeding. A somewhat larger dose would obviously be required to prevent such bleeding in every case.

A variation of this experiment is to give castrated female monkeys large doses of estrogenic hormone and then drop the dosage until bleeding sets in. A number of such experiments were reported by S. Zuckerman of Oxford, England, in 1936. This investigator found that whenever he reduced the daily dose from several hundred or several thousand international units to any amount below 200 international units, estrin-deprivation bleeding occurred thereafter. With more than 200 international units daily no bleeding occurred.

These observations suggest that the output of estrogenic hormone from the normal ovaries, which is of course sufficient to prevent estrin-deprivation bleeding, may be about 150 or 200 international units.

A third means of estimating the probable amount of estrogenic hormone produced by the monkey is provided by the fact that the so-called sex skin (the red swollen areas of the rump and thighs) is under the control of the hormone. Removal of the ovaries leads to shrinkage and pallor of the sex skin, whereas administration of an estrogen in sufficient amount will within a few days restore the color of this region. One of my animals, a young adult, had its ovaries removed at a time when its sex skin was in exceedingly florid condition. The whole area over the rump, the back of the thighs, and the ventral side of the base of the tail was swollen, thrown into distinct ridges and deep red in color. After the operation, but on the same day, the color and swelling were slightly reduced, owing no doubt to the stress of the operation. One hundred twenty-five international units of estrone was given daily beginning on the day of operation. A slight but definite decrease in the color of the sex skin took place during the next 10 days. The swelling of this region diminished also, although not to the same extent as the color, and on the 10th day, in spite of the administration of estrone, external vaginal bleeding began.

From the facts that the condition of the sex skin retrogressed very slowly and external bleeding did not begin until the 10th day, that is to say later than if the animal had been given no estrogen at all, it seems that the dose of 125 international units daily was almost but not quite sufficient to maintain the animal in the same condition as before removal of the ovaries. We may estimate therefore that 150 or 200 international units would have been required to substitute completely for the ovaries.

Another method of arriving at the desired information depends upon the fact that effective action of progesterone upon the endometrium to produce the progestational ("premenstrual") condition in castrated female monkeys requires that an estrogenic hormone be administered in conjunction with the hormone of the corpus luteum. Engle of Columbia University (1937) stated that a satisfactory combination for the castrated Rhesus monkey is 30 Allen-Doisy rat units of estrone (this is approximately 150 international units) daily with 0.5 Corner-Allen unit of progesterone (approximately 0.5 milligram). Hisaw and Greep (1938) gave a similar figure, i.e. about 150 international units of estrogen with 0.5 mg. of progesterone. These experiments are subject to the difficulty that they involve varying the dosage of two hormones simultaneously. The animals used by Hisaw and Greep were, moreover, relatively young, and the progestational changes involved were not as elaborate as the natural progestational state. The result fits however those obtained above in giving 150 international units of estrone as somewhat less than an adequate substitute for what the animal's own ovaries produce.

Probable daily output of estrogenic hormone by the monkey. By all these means of estimation we arrive at the result that the probable daily output of estrogenic hormone by the young adult Rhesus monkey is equivalent to somewhat more than 150 international units and may be set tentatively at the equivalent of 200 international units of estrone. This is a minimum figure; the true amount may be greater, but can hardly be smaller.

This estimate is based necessarily on the assumption that an ovarian hormone injected once daily in oil solution is utilized by the body as efficiently as hormone produced in the animal's own ovaries. There is at present no way of ascertaining how nearly this is true, but in any case, the slow absorption of an oily solution must afford a fairly good imitation of natural processes, and we may recall that my estimate of the rate of secretion of progesterone, which involved the same assumption, turned out to be near the truth, when checked by the recovery of the excretion product, pregnanediol.

Administration of estrogenic hormone in pellets. In recent years many experimental workers have been administering estrogenic hormones by compressing them into small hard pellets which are buried under the skin. This method has the advantage that the hormone is absorbed continuously from the surface of the pellet, whereas when given by injection the rate of absorption fluctuates, being high just after an injection and lower as more and more of the injected dose is absorbed. For this reason a given effect is obtained from a smaller daily dose when absorbed from a pellet than when injected. Pellets will probably be used in human cases in which long continued action is required, not only because of the continuous absorption, but also because insertion of the pellet, which can be done through a hollow needle, avoids repeated hypodermic punctures.

It becomes important for our present calculations to know just how much more effective this method is, dose for dose, than injections, because our estimate of the daily need for estrogenic hormone is based on comparison with injected doses. At first sight the results obtained by pellets seem to be achieved by very small doses indeed, and it is generally taken for granted that the method is much more effective than injection. Very little, however, is known about the exact comparison by actual experiment. Carl G. Hartman has reported, for instance, that the sex skin of a Rhesus monkey was kept red and swollen for 4 months with a single 3-milligram pellet of estrone. This seems small, but assuming that the pellet was entirely used up, this gives a daily dose of 0.025 mg. or 250 international units (3 mg. divided by 120 days). Hartman's monkey thus actually received a larger daily dose than is called for on the basis of our estimate, namely 200 international units.

Deanesly and Parkes of London, who introduced the pellet method, cite an experiment with one of the male hormones which indicates that the dose by pellet is about one-half that by injection. I have heard of experiments with another male hormone in which the ratio was 2 to 3. If any such proportion as these is true of estrone, then our calculated daily output of estrogen by the monkey is roughly one-third to one-half too large.

But here again we are guessing. What evidence is there that absorption from a pellet is really comparable to the natural absorption of the animal's own hormone from the ovarian cells? Nobody knows the answer to this. It might even be true that the pellets yield their substance to the blood stream more easily than do the cells of endocrine glands, in which the hormone is made and stored within the cellular substance. The big molecules of the hormone have to make their way through the outer layers of the cell, not merely drop off the surface of a pellet.

In short, it would be premature to let the pellet method outweigh, in these very crude calculations, the results from injections, about which at present we know very much more; and therefore I propose to maintain for the present the estimate arrived at above, namely that a female Rhesus monkey produces from her two ovaries estrogenic hormone equivalent to about 200 international units of estrone daily.

If this were actually estrone, the daily output would weigh 0.02 milligram.

Application of these calculations to the human female. Can we apply this estimate to the human female? A woman weighs on the average 15 times as much as a monkey. Two hundred international units x 15 gives 3,000 international units as the daily output of the woman's ovaries. Various medical observations which have been published might be analyzed to give us some sort of check on the estimate. My friend, W. M. Allen, for example, informs me that in his treatment of women whose ovaries had been removed previously he could induce menstruation-like bleeding with estrogenic hormone equivalent to about 4,200 international units of estrone daily. This figure, which is the most pertinent I can find, is roughly 50 per cent higher than my calculation from the monkey experiments, but we have no way of telling how closely this amount compares with that needed by a normal woman. Perhaps what Allen was trying to do required more, perhaps less, than the normal output.

Unfortunately we cannot get help in this problem by measuring the estrogen discharged in the urine, because we do not know just what relationship exists between the hormone which is at work in the body and that which is excreted. The number of milligrams of estriol and similar substances recovered from the urine does not tell us how much of the ovarian hormone was made and used in the body (Appendix II, note 16).

This chapter has contained, after all, a good deal of valiant shooting in the dark. Our conclusions and estimates are sure to be better in time to come. Meanwhile the reader will perceive that after the excitement and drama of the pioneer phase of research on the ovarian hormones, we are in for a lot of tedious, unspectacular measurement and computation, until the reactions of these substances in the body are quantitatively known as well as the chemist knows the reactions in his flasks. Then, as Lord Kelvin said, we shall really understand our subject.

THE HORMONES IN PREGNANCY

"Long was I hugg'd close—long and long.
Immense have been the preparations for me,
Faithful and friendly the arms that have help'd me.
Cycles ferried my cradle, rowing and rowing like friendly
        boatmen,
For room to me stars kept aside in their own rings,
They sent influences to look after what was to hold me.
Before I was born out of my mother generations guided me,
My embryo has never been torpid, nothing could overlay it."
              —WALT WHITMAN, *Song of Myself.*

CHAPTER VIII

THE HORMONES IN PREGNANCY

THE maintenance of pregnancy is a truly complex affair. A living creature is growing at a tremendous rate inside a hollow chamber, the uterus. This organ must at first tolerate, even support the newcomer. It must grow in size and strength so that its enterprising tenant may not overwhelm it (Fig. 28). All the other muscle-walled organs of the body are built to keep things moving—the heart, the intestines, the bladder for example—and so, ultimately, is the uterus. For nine months, however, it must be kept in check and not allowed to expel the infant prematurely. Then all of a sudden its energies are released and it is called upon to deliver its contents into the world, through the narrow bony canal of the pelvis, with sufficient force and speed on one hand, and sufficient gentleness on the other, to avoid wearing out the mother or crushing the baby. To use a current expression of bewilderment, figure out all that if you can! Nature, indeed, has figured it out reasonably well; but when the physiologist attempts to discern the factors of this multifarious process and to see how they are set in motion, timed, and controlled, he finds he has yet a long research ahead of him.

In this book we can do no more than sketch the problems involved. In outline, what has to be worked out is the growth and function of a muscular organ, controlled in part by the nervous system, in part by hormones. The latter are those which come from the ovary, the pituitary and the adrenal, together with the output of a new source, peculiar to pregnancy, the placenta.

THE PLACENTA

Once the embryo is safely lodged in the uterus and has begun to grow, a new era of hormone activity begins. The

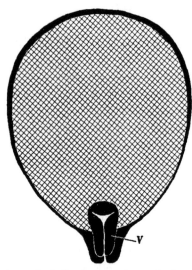

Fɪɢ. 28. Enlargement of the human uterus during pregnancy. At *v*, virginal uterus; the large hatched area is the pregnant uterus, at full term, drawn to same scale, approximately ¼ natural size. Adapted from a figure by Stieve.

task of the ovary is not yet over, but its functions are in large part to be taken up, reinforced, and even superseded by a new organ of internal secretion, the placenta. The human placenta is an object of considerable size, as shown in the fine old engraving from Casserius (Plate XX). When fully developed it is about 18 centimeters (7.5 inches) in diameter, and weighs 500 grams (a little more than one pound) on the average. When in place in the uterus, its structure is like nothing else so much as the matted roots of a tree planted in a tub. The roots simulate the villi of the placenta, which carry within their slender strands the finest branches of the blood vessels coming from the infant in the umbilical cord. Through these delicate vessels the blood of the infant circulates in a constant and rapid stream. The tub in which we imagine the tree planted simulates the pool which the embryo has excavated for itself in the wall of the mother's uterus (Fig. 14, p. 58).

Through this pool the mother's blood slowly flows, out from the arteries that supply the region and back into the veins, bathing the rootlike villi of the placenta. From the mother's blood to the blood of the embryo, oxygen and dissolved foodstuffs filter through the covering cells of the villi and through the thin walls of the blood vessels that run along within them, just as nutritive substances pass into the roots of a plant from the moist soil in which they grow. The infant sends back carbon dioxide, urea, and other wastes from its tissues, to be filtered out through the placental vessels into the mother's blood, which carries away these wastes to be disposed of with the products of her own metabolism.

The covering cells of the villi are called upon, however, not only to take part in the process of filtering foodstuffs inward and waste products out, but also to produce a whole series of endocrine substances. This important fact is unfortunately masked by a great deal of confusion in our present knowledge. The placenta differs greatly in different species, not only in its structure but also in its endocrine activities. Statements which are true of one species may not apply at all in others. For this reason our discussion will be limited to the human species (except as specifically stated) and even then we shall be restrained by a certain amount of uncertainty. To begin with, the placenta begins very early in pregnancy to elaborate a hormone of its own, having profound gonadotrophic properties (that is, power to stimulate activity of the ovary and the testes) much like that of the pituitary gonadotrophic hormones.[1] This activity begins, indeed, before we can properly speak of the placenta, for the hormone in question is made by the cells (i.e. the trophoblast) covering the early villous processes that surround the embryo, as soon as they

[1] For the sake of clearness, it seems best to refer to the gonadotrophic material in the singular, i.e. "a hormone," but actually it seems to be a hormone complex comprising two substances, one of which tends to stimulate growth of the follicles, the other to convert the follicles into corpora lutea (Appendix II, note 11).

begin to grow. Knowledge of this subject goes back to 1928, when Aschheim and Zondek, then of Berlin, found that during pregnancy the human urine contains something that has a powerfully stimulating effect on the ovaries of young mice and rats. This provided the basis for the now famous Aschheim-Zondek test for pregnancy. The urine of a pregnant woman, injected into an infantile mouse or rat, produces prompt and characteristic signs of activity in the ovarian follicles. An even quicker test for pregnancy is provided by a modification of this procedure, introduced by Maurice Friedman, an American investigator. In the Friedman test the urine, either raw or partially purified by precipitation with alcohol, is injected into the ear vein of a rabbit. If the patient is pregnant, the rabbit ovulates about 10 hours after the injection. This hormone test for pregnancy is (in both variations) probably more nearly infallible than any other biological test used by physicians, for when properly performed it is accurate in better than 98 per cent of all cases. '

The gonadotrophic hormone complex of the urine can also be extracted from the placenta and in all probability is made there. George O. Gey, a tissue culture expert of Baltimore, recently showed that placental tissue growing in his test tubes was able to produce a gonadotrophic substance. Chemically the urinary gonadotrophic material is protein, like the gonadotrophic hormone that is produced in the pituitary gland, and indeed it is so much like the latter that it was for a time considered to be identical with it, but clear differences between the two substances have been observed, as evidenced by the details of their effects on animals of various species. The placental gonadotrophic hormone complex appears in the urine in the first month of pregnancy, in sufficient amount to give a positive Aschheim-Zondek or Friedman test. It is present throughout pregnancy, but reaches its greatest amount in the second month and falls off rapidly thereafter. In the Rhesus monkey it is found only between the 18th and the 25th day. In the blood of the pregnant mare there is a gonadotrophic hormone which also has a stimulating effect on the ovaries of test animals, such as rats, differing considerably, however, in detail. This does not get into the urine in significant amounts, which means that it must be different chemically from the human hormone of similar action.

We do not know why these substances can be found in some species of animals and not in others, nor do we know what function they perform in pregnancy. The whole series of pituitary and pituitary-like hormones has been extremely difficult to investigate chemically because the substances are proteins and they defy purification. The ovaries of the rat and the rabbit can distinguish them better than the chemist. For the present we must content ourselves with being grateful for the pregnancy test and await the day when these troublesome substances yield themselves to chemical isolation.

As mentioned previously (Chapter IV) the urine of pregnant women contains relatively large amounts of estrogenic substances, which increase as pregnancy advances and disappear after parturition. These substances have been found in the urine of several other species during pregnancy. The human placenta also contains large amounts of estrogenic hormones, chiefly estriol, and is almost certainly the source of those which appear in the urine. As mentioned in Chapter IV, when the ovaries are removed during pregnancy, estrogens continue to be excreted in the urine, a fact which proves that some other source exists, and this can hardly be anything else than the placenta. A similar situation, produced experimentally in the monkey, has been studied very carefully and reported by R. L. Dorfman and Gertrude Van Wagenen of Yale Medical School.

There are several possible ways in which the production of estrogens by the placenta may be useful. It has been suggested that these hormones are needed, in larger amounts than the ovaries can provide (a) to promote the growth of the uterus which occurs during pregnancy; (b) to cause growth of the mammary glands to get them ready for the production of milk; (c) to set up contractions of the uterus when the time comes for parturition; (d) to cause persistence of the corpus luteum. These possibilities will be discussed again later in this chapter.

As regards progesterone in the placenta, it has already been pointed out that in some animals the corpora lutea are indispensable until the end of pregnancy. In the rat and the cow, for example, the ovaries cannot be removed at any time without causing loss of the embryos, and the corpora lutea appear to be functional until almost the end of gestation. In the human species, on the other hand, the ovaries can be removed without affecting the survival and birth of the infant, as early as the third month of pregnancy. After such an operation pregnanediol has been found in the urine, an indication that progesterone has continued to be produced somewhere. Naturally the placenta has been suspected, and assays have yielded small amounts of progesterone or a substance of closely similar kind.

ACTION OF THE HORMONES IN PREGNANCY

We can do hardly more here than sketch what is known about the multifarious interactions of the hormones in pregnancy. Readers who wish to follow the subject in detail may consult the excellent book on this subject by my colleague S. R. M. Reynolds.[2] In the first place, the two ovarian hormones contribute to the growth of the uterus. Growth of the muscular wall is known to involve (as one might well expect) first an increase of the number of the muscle cells, and then an increase of the size of the individual cells. In the human uterus the measurements of the German histologist Stieve indicate that the muscle cells are 17 to 40 times larger at the end of

[2] *Physiology of the Uterus, with Clinical Correlations,* by Samuel R. M. Reynolds, New York, 1939.

pregnancy than in the empty uterus. Other elements of the uterine wall, namely the connective tissue, blood vessels, and nerves, also increase in amount. In this process the estrogenic hormone contributes its general growth-promoting effect, which it exerts by augmenting the blood flow through the tissues of the uterus. Under its influence there is some increase in number of the uterine cells. Progesterone in turn causes a decided wave of cell division in the muscle, and then within a few days at the beginning of pregnancy there is a large increase in the number of muscle cells. Subsequent growth of the individual cells, and consequently of the whole wall of the uterus, comes about as the result of stretching by the growing embryo. As the uterus is distended it grows in thickness and strength; if this were not so, the infant would soon rupture the walls that confine it. Everyone is, of course, familiar with the fact that working a muscle makes it grow, and this is no less true of the involuntary muscles of the internal organs than of the skeletal muscle; but like most other familiar responses of the body, we often take it for granted without realizing how little we know how it comes about. Why the uterine muscle grows when that organ is distended is a large question of general physiology, beyond the scope of this book. Reynolds has shown that if he distends the rabbit's uterus by introducing pellets of wax, it will grow in thickness just as it does in pregnancy. By this means he has been able to test the effects of the ovarian hormones upon the growth-response to distention, and has found (among many other interesting facts) that treatment with estrogenic hormone cuts down this response. This hormone, then, which at first helps start the growth of the pregnant uterus, afterward helps to control it. In human pregnancy we know there is plenty of estrogen available in the later months; in all probability this serves to keep the growth of the uterus from going too far. With this hint that the interplay of the hormones is indeed complex, we had better leave the subject to the specialists for further study.

Finally the time comes when gestation can go on no longer. The uterus, overburdened by its rapidly growing tenant, must deliver itself. Degenerative changes begin in the placenta, the nourishment of the infant is thereby impaired, and the uterus commences its efforts to expel the child. This it accomplishes by means of strong contractions, efficiently timed and coordinated so that the infant is pushed toward the outlet of the uterus. The uterine orifice, and afterward the vaginal canal, are stretched open to allow passage of the infant, while the rest of the uterus contracts to provide the necessary force. There is no simple explanation of the onset of labor. Many physicians and biologists have tried to discover some simple reason why at a particular time—9 months in the human, 2 years in the elephant, 21 days in the mouse, or whenever, according to the species, the fated hour arrives—the act of parturition begins. It has been thought, for example, that the uterus is at last simply stretched too far and is thereby irritated into contracting again; or that the breakdown of the placenta constitutes a stimulus to the uterine muscle; or that some chemical substance from elsewhere in the body sets the muscle into action. When it was discovered that estrogenic hormones stimulate the involuntary muscle of the uterus, and that progesterone tends to relax it, an attractive theory of the cause of labor at once suggested itself. We need only suppose that when the end of gestation draws near, the production of progesterone goes down, and estrogenic hormone is thereby allowed to build up contractions of the uterine muscle. This hypothesis is however much too simple, as Reynolds points out in the book cited above. For one thing, the contractions of the uterus in labor are very different in their timing and coordination from those of the nonpregnant uterus. The fact is that the uterus at the end of pregnancy is operated by a very elaborately organized set of adjustments.

The proportion of the infant to the space it occupies; the strength of the uterine wall and the pressure it exerts upon its contents; the rate of blood flow through the uterus; the sensitivity of its muscle and nerves; the balance of the hormones that affect it; the nutrition of the infant and the placenta—all these factors (and others beside them, for all we know) are balanced one against the other and when the crisis comes they are all involved at once. The physiologist who looks for one specific cause of the onset of labor is up against the same kind of problem as the economist who tries to find one single cause for a stock market crash or to pin down a nationwide problem of unemployment to one specific factor. When dealing with such complex affairs as those of a nation or a pregnancy the investigator cannot isolate one factor at a time and study it singly. He has to unravel a whole system of balanced forces. In the problem we are considering, the hormones are certainly to be numbered among the most important factors, but it is scarcely safe at present to say more.

LACTATION

When the mother's body has completed its provision of shelter and nourishment for the child by means of the hormones and has seen him safely into the world, it has yet another service to render on his behalf—namely that the breast for which he will so promptly cry is ready to supply him with milk.

The recent discovery of a specific hormone for lactation, in the pituitary gland, was a great surprise. It involved a simple little piece of scientific logic which the reader may enjoy after the preceding complexities of this chapter. We had better clear the way for this story, however, by recalling to mind the earlier history of the mammary gland. When a girl or a young animal reaches sexual maturity, the mammary glands are brought from the immature state to the adult condition, by action of the estrogenic hormone. Nothing is more striking than to watch the growth of the mammary glands in a young animal receiving estrogen. Each gland is a system of branching channels lined by cells derived from the outer layer of the skin (epidermis). Long before birth these ducts begin to grow from the nipples and to spread out around them in a little circle under the skin. At first the channels are few and short, and only slightly branched (Fig. 29, A). Under the action of the estrogenic hormone they branch extensively and spread to adult dimensions (as indicated in the diagram, Fig. 29, B). In the girl at puberty this is, of course, a gradual change, but in an experimental animal under hormone treatment it can be produced quite rapidly. The nipples quadruple in size in a few days, and the ducts push outward in a widening circle. In pregnancy a much greater development occurs. The branches of the duct system develop extensive terminal twigs ending in secretory alveoli (Fig. 29, C, D). These become more numerous as pregnancy advances. Finally, globules of milk-fat accumulate in the cells of the alveoli (Fig. 29, E). The actual flow of milk in quantity does not begin, however, until after parturition.

From the time it was first conjectured that the corpus luteum is a gland of internal secretion, until quite recently, it was supposed that this particular endocrine organ is responsible for the growth of the mammary gland in pregnancy and the secretion of milk. On the face of it, there could hardly be a more plausible conjecture, for the growth of the mammary gland closely follows the appearance of the corpus luteum, and is so obviously a part of the general preparation for the infant that it seems very logically to go with the other functions that the ovary exerts during pregnancy.

If this is true, however, why does not the corpus luteum produce mammary growth and even lactation not only in pregnancy but in each ovarian cycle? To this query A. S. Parkes of London in 1929 offered the tentative reply that

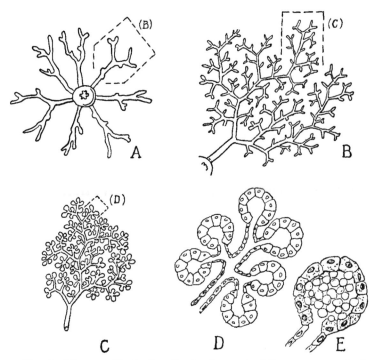

FIG. 29. Diagrams illustrating the development of the mammary gland as seen in laboratory animals. *A*, mammary gland of immature animal, consisting of simple ducts radiating from the nipple. *B*, small area of *A* enlarged to show adult virginal gland. The action of estrogenic hormone has produced extensive growth and branching of the ducts. *C*, small part of *B* enlarged again to show the effect of pregnancy; there has been a great development of duct twigs with terminal alveoli. *D*, terminal alveoli enlarged to show their cell-structure. *E*, secretion of milk globules by the cells of the alveoli. Based on a figure by C. W. Turner in *Sex and Internal Secretions*.

perhaps in the ordinary cycle, in which the corpus luteum lasts but two weeks, there is not time to build up the mammary gland sufficiently. He therefore ingeniously proposed to apply the discovery, then quite new, that the corpora lutea can be made to persist for weeks by injection of anterior pituitary extract. He tried this experiment in rabbits, by mating them to males rendered infertile by having their seminal ducts blocked, so that these females ovulated but did not become pregnant. The corpora lutea, which under these circumstances would normally degenerate after about 14 days, were made to persist by the pituitary hormones. The expectations of the experimenter were brilliantly fulfilled by the occurrence of mammary growth and lactation. When the report of these experiments reached the United States, we had in our University of Rochester laboratory a fair amount of the then new corpus luteum hormone in crude form (progestin) and took the obvious step of trying to produce mammary growth and lactation with the progestin alone. To my surprise we got no lactation and no marked growth of the mammary gland. From this failure, however, an important deduction could be made. Parkes had subjected his rabbits to the action of two hormones (pituitary and corpus luteum); I had given only the latter. He obtained lactation, I did not. Subtracting my procedure from his, it looked as if the anterior pituitary hormone was the only necessary factor. We made up a flask of pituitary extract from sheep and injected it into six adult castrated virgin female rabbits. In 3 days milk began to drip from their nipples; in 10 days the mammary glands had thickened and spread all over the chest and belly as in advanced pregnancy and when we milked them the milk spurted across the room.

I thought at first that this discovery of the lactogenic potency of the anterior lobe of the pituitary was completely new, but study of the scientific journals revealed that a few months previously two Alsatians, Stricker and Grueter, working in Strasbourg in the laboratory of Paul Bouin, had done exactly the same experiment with the same result. By what process of logic they were led to try it was not narrated in their report.

The extracts were very crude and we set out to purify them. One of the large drug houses provided more extract by the quart and one of their research staff started to work out the chemical separations, but he and I found only that we were dealing with a protein, after which we simply got lost in this most difficult field of biological chemistry. Meanwhile my genial and versatile friend Oscar Riddle of the Carnegie Institution (Department of Genetics, at Cold Spring Harbor), ably assisted by his colleague R. W. Bates, applied his all-round knowledge of tissue chemistry to the task and succeeded in purifying the hormone to a considerable degree. He called it *prolactin*. The complete chemical isolation and chemical identification of this important substance is now a problem for the most advanced special experts in the chemistry of proteins (Appendix II, note 17).

There is a queer sequel of this discovery of prolactin, which opens a vista of the long past origin of the hormones in evolutionary history. This story has to do with pigeons' milk. Not that pigeons have mammary glands; but females of the genus actually secrete into their crops a kind of milky secretion which they regurgitate and feed to their nestlings. This crop-milk is produced by special glands in the lining of the crop. Dr. Riddle knows all about pigeons; he has been studying their physiology for years and had a fine collection of them at Cold Spring Harbor. Impressed by the parallelism between the formation of crop-milk and mammalian lactation, he administered his extract of beef pituitaries to some of his female pigeons and got proliferation of the crop glands just as if they were mammary glands. This reaction is so easy to produce that it is now the standard test for prolactin. The strangest part is, however, yet to be told. The eggs of the amphibians (frogs, toads, and salamanders for example) are laid in the water and the embryos have the benefit neither of nest and crop-milk nor of uterus and mammary gland. When the eggs are shed by the mother, however, they are protected by an envelope of jelly, laid on in the mother's oviduct. An Argentine physician, already mentioned in this work, Dr. Inés de Allende of Córdoba, has discovered that a hormone

like that found in extracts of beef pituitary glands is responsible for the secretion of the protective jelly of the eggs of
toads. She elicited this function by inserting pieces of toad
or beef pituitary gland under the skin of her toads, thus increasing the available amount of pituitary hormones; and
in a few cases she could even elicit it by injecting Riddle's
prolactin.

Here, then, are three particular means of provision for the
newborn infant, occurring in three widely different branches
of the animal kingdom, and adapted to offspring living under
very dissimilar circumstances, yet all these secretions are controlled by the pituitary gland and can be induced by extracts
of beef pituitaries. The embryologist perceives a further
remarkable feature of this story. The mammary gland, he
knows, is derived from the outer of the three fundamental
tissue layers of the embryo, the *ectoderm*; the crop glands
of the pigeon are derived from the inner layer, the *endoderm*;
and the secretory lining of the toad's oviduct is derived from
the middle layer, the *mesoderm*. These three tissues from which
the pituitary gland elicits reactions of such similar usefulness, are about as widely different in their position in the
body, and in their embryological history, as they can possibly
be, but when in the evolution of toad, bird, and mammal there
was need to call upon them to foster the fledglings of their
. kind, the pituitary gland took control in each case.

Has the corpus luteum, then, no role whatever in the processes leading to lactation, and can the pituitary extracts induce lactation in a mammary gland prepared only by estrogenic hormone? My rabbits seemed to indicate that this is
true, but Stricker and Grueter declared that their pituitary
extracts were not successful unless there had been corpora
lutea in the ovaries at some time during recent months; in
other words, the pituitary lactogenic hormone appeared able
only to act upon a mammary gland sensitized by progesterone. This difference in the findings led to a great deal of

subsequent work by various investigators, well summarized by C. W. Turner of Missouri, himself one of the leaders in this work, in *Sex and Internal Secretions*.[3] It now appears that the second of the three more or less distinct stages of growth and function of the mammary gland, already referred to and illustrated (Fig. 29), involves somewhat different responses to the hormones in different species. The first stage, that of preliminary growth of the duct system to the adult virginal state, occurs under the influence of estrogenic hormone as already described. In the second stage, the ends of the ducts proliferate and branch into numerous terminal alveoli. This stage in many animals requires the action of progesterone; in a few species (of which the guinea pig is an example) progesterone is not needed at all and the proliferation can be completely induced by the estrogens; in some other species growth of the alveoli is at least facilitated or speeded up by progesterone, though not actually dependent upon that hormone. Judging from various observations which have been reported on monkeys, the primate mammary gland is among this intermediate group. Whether this is also true of the human we do not know at present. The third stage, that of secretion of milk, is brought about by the lactogenic hormone of the pituitary. It is a striking evidence of the potency of these hormones to induce lactation that the rudimentary mammary glands of male animals can readily be made to lactate by a suitable course of treatment with the estrogen-prolactin or estrin-progesterone-prolactin sequence.

We have yet to discover how the pituitary gland is stimulated to exert its lactogenic effect during pregnancy. The reason the flow of milk does not begin until just after parturition, and then begins suddenly, is that lactation is inhibited by the estrogenic hormone of the placenta. Once the placenta is out of the way, the flow of milk is released.

[3] *Sex and Internal Secretions*, edited by Edgar Allen, Baltimore, 2d ed., 1939; Chapter XI, *The Mammary Glands*, by C. W. Turner.

The induction of lactation in pregnancy is thus another example of those remarkable integrative linkages among the various parts of the reproductive system by means of hormones, by which shelter, warmth, and nourishment are provided for the mammalian egg while it is incubated within the body and at the breast of its mother. By such extraordinary means as this have the children of men, latest of a long series of creatures to haunt the earth, been protected from isolation and danger in their earliest days, and given time to grow for nine months before being exposed to the rigors of the outer world. This privilege of uterogestation, which would be impossible without the ovarian and placental hormones, gives an incalculable advantage to mankind and the other mammals in the struggle for superiority in the animal kingdom.

THE MALE HORMONE

"She [Nature] spawneth men as mallows fresh,
 Hero and maiden, flesh of her flesh;
 She drugs her water and her wheat
 With the flavors she finds meet,
 And gives them what to drink and eat;
 And having thus their bread and growth,
 They do her bidding, nothing loath."

—RALPH WALDO EMERSON, *Nature* II.

CHAPTER IX

THE MALE HORMONE

THE male sex glands (testes) of man and the other mammals, like the ovaries, perform a double task. They exist primarily, of course, to produce the male germ cells. In primitive aquatic animals this is all they need to do. The Hydra, for example, shown in Plate II, *C*, in the act of discharging its sperm cells directly into the water, has fulfilled its reproductive task for the season, and its empty testes are of no more consequence than a spent skyrocket. In mammals, however, things are not so simple. There are other needs that can be fulfilled only by the coordination of various parts of the body by means of a hormone. Not only must the sperm cells be formed and ripened; they must also be stored until they are needed in mating. What is more, they must be stored in a most particular way, immersed in a watery environment, for the mammals have never fully shaken off their ancient adaptation to the sea. They spend their lives on land, but when the time comes to reproduce their kind, their spawning requires salt water—not indeed the actual sea, but the internal fluids of the generative organs. The egg ripens in the fluid of the Graafian follicle. The sperm cells accomplish their tortuous journey to the egg by swimming, and the offspring of all the mammals spend the long term of gestation in a submarine environment. You and I cannot remember our ancestral life in the water, nor the nine months we ourselves lived beneath the chorio-amniotic sea, but our tissues recall it; the skin, the kidneys and the adrenal glands working to hold sufficient water and just enough salt, the testes providing through their accessory organs those fluids in which the sperm cells may be effectually launched upon the sea of life.

In another way also the endocrine function of the testis becomes necessary. The higher animals lead complicated lives.

They wage a varied battle for existence and are swayed by many circumstances. Perpetuation of the race amid such distractions requires especially active maintenance of sexual vigor and the urge to mate. This too becomes a function of the testis as an organ of generation.

THE MALE REPRODUCTIVE ORGANS

The testis. To follow the story in detail we must first review the anatomy of the testis and the associated male organs of reproduction. The diagram (Fig. 30) will serve to orient us. The two testes lie in their pouch of skin (*scrotum*). They are objects of ovoid shape, about the size of walnuts, that is to say 4.5 x 3 centimeters (2 x 1¼ inches) in diameter. Within these small rounded bodies the business of sperm cell production is carried on inside an extraordinary system of tubular canals. The testis, in fact, consists essentially of many hundred small tubules, called *seminiferous tubules*, each about 30 to 70 centimeters (1 to 2 feet) in length and about as large in diameter as a strand of sewing silk. These tubules are coiled very tightly in the small space available, and thus when we look through the microscope at a section of the testis, we see countless sections of the individual tubules, cut in every possible direction (Plate XXIII, *A*). How they actually run, and how they are connected, was for a long time one of the most difficult problems of microscopic anatomy. We should get a similar picture if we took a thoroughly tangled ball of twine and cut a section through it with a sharp knife. Nobody could possibly tell from such a cut whether the ball of twine contained one long piece of twine, or several shorter ones, or how they were joined together. Likewise a section of the testis cannot tell us anything about the course of the seminiferous tubules. For a century microscopists applied their various technical tricks to this problem, including the making of magnified models from serial sections, but with only imperfect success, owing to the difficulty and laboriousness of following

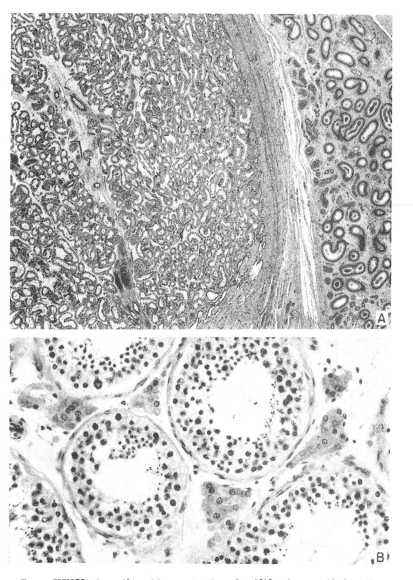

PLATE XXIII. *A*, portion of human testis and epididymis, magnified 10 times. The large circles and loops at the right are sections of the coiled tube of the epididymis; the smaller tubules filling the left two-thirds of the picture are the seminiferous tubules of the testis. *B*, area of the same testis magnified 200 times, showing 5 clumps of interstitial cells between the tubules. Photographs from preparation lent by Joseph Gillman through I. Gersh.

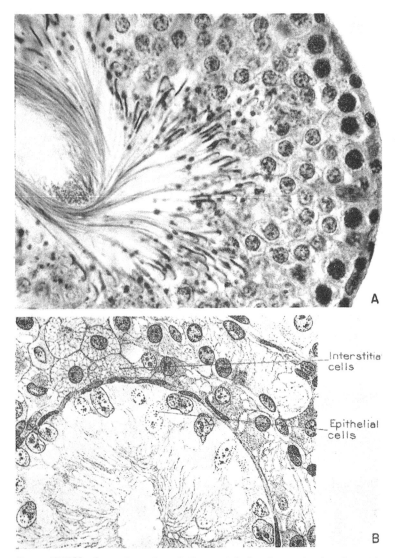

Interstitial
cells

Epithelial
cells

A

B

PLATE XXIV. *A*, portion of seminiferous tubule of rat, showing formation
of sperm cells. Note the peculiar hook- or halberd-shaped heads and long
tails of the rat's sperm cells (form of human sperm cells shown in Fig. 7).
Magnified 600 times. From specimen lent by K. E. Mason. *B*, portion of seminiferous tubule of undescended (cryptorchid) testis of pig. Note that sperm
cell formation is totally absent, the tubule being lined by ill-defined cells
(*epithelial cells*). The interstitial cells are, however, well preserved. Magnified
about 600 times.

these minute and lengthy canals. Finally in 1913 the problem was cleverly solved by the late Professor Carl Huber of the University of Michigan, who worked out a method of soften-

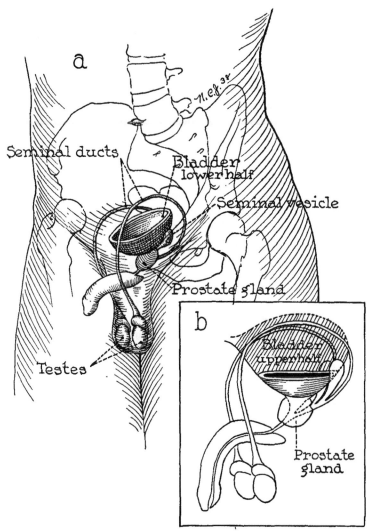

Fig. 30. The human male reproductive system. From *Attaining Manhood*, by George W. Corner, by courtesy of Harper and Brothers.

ing the testis with acid and then, with incredible patience and dexterity, dissected out complete tubules under the microscope with fine needles and mounted them, without breakage, on glass slides. He found that they are arranged in loops or arches, all opening into a network of channels at the hilum of the testis, whence they are drained off by a dozen larger channels into the *epididymis*. All these facts have been depicted in Fig. 31.

In cross section under high magnification (Plate XXIV, *A*) the tubules are seen to be lined by several layers of cells. The outermost layer (i.e. that farthest from the central channel of the tubule) is made up of large clear cells. As these divide to form the next and succeeding layers, the cells become smaller. Finally little is left except the nucleus, and even this becomes more compact. A long tail-like process grows out of the rapidly shrinking cell. This figure represents the rat's sperm cells with their hook or halberd-shaped heads. The completed human sperm cell is shown in Fig. 7. It has a total length, including the tail, of about 60 microns or 1/400 inch. The head is about 5 microns long by 3 wide. The sperm cell is therefore by far the smallest cell in the body. An idea of its relative dimensions may be gained by comparing it with the printed period at the end of this sentence. If the sperm heads were laid like a pavement, one layer deep, on such a dot, it would take about 2,500 of them to cover it. About 12 egg cells could be placed on such an area.

As the sperm cells are formed, little clumps of them cling to supporting cells in the lining of the tubules until finally they drop off and are carried along the channel toward the seminal ducts. In animals that breed at all seasons of the year, for instance man, rat, rabbit and guinea pig, sperm production goes on continuously, passing in waves along the tubules, so that sperm cells are always available. Many wild species, however, have distinct breeding seasons once a year, and in these there is a cycle of testicular activity. In

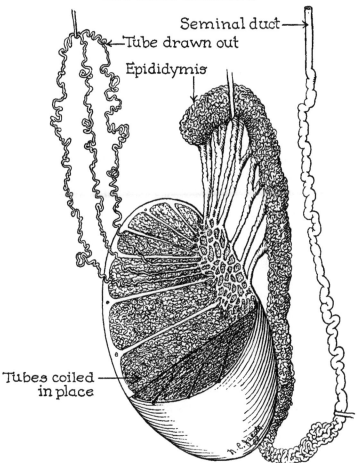

Seminal duct

Tube drawn out

Epididymis

Tubes coiled
in place

Fig. 31. Arrangement of the tubules of the testis, the epididymis, and the seminal duct. Adapted from Spalteholz and Huber. From *Attaining Manhood*, by George W. Corner, by courtesy of Harper and Brothers.

the seasons of inactivity spermatogenesis ceases and the testis diminishes in size.

The testis, like the ovary, is under control of the pituitary gonadotrophic hormones. One of the most striking results of the brilliant campaign of investigation of the pituitary led by Philip E. Smith, now of Columbia University, has been the discovery (1927) that removal of the pituitary gland from an adult male causes degeneration of the testis, and in an immature male prevents the development of sperm cell formation. The implantation of bits of pituitary gland or the injection of pituitary extracts will substitute for the missing organ and prevent degeneration of the testis. In some species, but not in others, success has been attained in starting sperm formation in immature animals by injecting pituitary extracts.

Another remarkable discovery about spermatogenesis was announced by Carl R. Moore (University of Chicago) in 1924, namely, that the mammalian testis cannot form sperm cells unless it is subjected to a temperature slightly lower than that of the interior of the body.[1] In its normal place in the scrotal sac the testis is under temperature conditions exactly suited to its function. It has long been known that a man with undescended testicles is not fertile, and horse breeders are well aware that the same is true of cryptorchid stallions. They often call upon veterinary surgeons to bring down the testes of colts by operation. Moore's investigations have given us the reason for this sterility. In man descent of the testes normally occurs before birth. The sex glands are formed in the abdominal cavity near the lower pole of the kidneys, but they gradually move downward (or as some embryologists prefer to say, the body grows upward past the testes) until they have fully descended and have occupied their permanent place in the scrotum. Just why this elaborate transfer of the testes takes place, seemingly leaving them less protected than if they remained in the abdominal cavity, as in lower vertebrate animals, has never been satisfactorily explained. One can only guess that when the process of evolu-

[1] For a full account of this subject, see *Sex and Internal Secretions*, edited by Edgar Allen, 2d ed., Baltimore, 1939; Chapter VII, "Biology of the Testes," by Carl R. Moore.

tion of the warm-blooded condition was accomplished, Nature discovered (so to speak) that after all the testes could not stand the temperature at which she had stabilized the mammals, and had to get them to a cooler place; but any such conjecture seems to put Nature (or whatever you choose to call the forces that guide evolution) in a position like that of tariff legislators, chess players and others who find that one change in a complicated situation may set off unexpected changes elsewhere. At any rate, descent of the testes is a deep-seated phenomenon that has become essential to fertility. The inside of the scrotum is several degrees cooler than the abdominal cavity, because the scrotal sac has thin walls and no insulating layer of subcutaneous fat, but numerous sweat glands by which it loses heat. Moore found that if he put the testis of a guinea pig or other animal back in the abdominal cavity (preserving its blood supply) and stitched it there, the seminiferous tubules became disorganized within a week, but recovered if he brought them down again before the damage had become permanent. He took also a fertile ram and wrapped its scrotum in woolen coverings so that the testes were brought to the temperature of the rest of the body. This too caused cessation of sperm production. Actual direct heating of the testis has a similar effect. A single exposure of the guinea pig's testis for 15 minutes to a temperature 6 degrees above that of the interior of the body causes degeneration of the seminiferous tubules. It is known that in man high fevers are followed by temporary loss of spermatozoa.

Descent of the testis into the scrotum is part of the general pattern of the development of the sex gland and is therefore subject to control by the pituitary and pituitary-like hormones. It has been known for about 10 years that gonadotrophic hormones from pregnancy urine can be used for the treatment of non-descent of the testis in boys. It does not always succeed, for adhesions and other obstacles may interfere; therefore the treatment must be used only in selected cases after thorough study. When the hormone fails, surgical methods are usually still available.

Interstitial cells. In the angular spaces between the tubules the microscope reveals little clumps of relatively large cells not in any way connected with the sperm-forming cells (Plate XXIII, *B*). Although there are only a few cells at any one point, there are so many clumps that the total mass of the interstitial cells adds up to a significant proportion of the whole testis. Some writers call the totality of these cells "the interstitial gland." There is an ample network of capillary blood vessels among these cell clumps. The arrangement is obviously like that seen in the glands of internal secretion; and it is in fact very probable that this is the source of the testicular hormone. We cannot however be sure that the hormone is not made by the cells of the seminiferous tubules, as will be discussed later.

The genital duct system. When the sperm cells have reached completion in the testis, they are perfectly formed but inactive. Freed from their parent cells, they are swept passively along the tubules into the larger ducts that drain the tubule system. If the sperm cells were discharged from the body in this nonmotile state they could not fulfill their task of reaching and fertilizing the egg cell. They require a further period of ripening and conditioning until they become fully motile and potent. Furthermore the seminal fluid in which they are to be carried must be made and added to them, bringing suitable substances for their nourishment and stimulation.

These needs are served by a complex system of ducts and accessory glandular structures. The dozen or more ducts that leave the testis all drain into a single tube about 7 meters (21 feet) long, which is tightly coiled, as shown in Fig. 31, into a dense mass, the *epididymis*, which lies upon the testis. This coiled duct is lined by special secretory cells and is believed to function as a storage place for sperm cells, which take many days to be carried through its whole length. During this journey they have time to mature. If one examines under the microscope sperm cells taken from the epididymis of a freshly killed animal, they are found to be in active motion, whereas those taken from the testis are motionless. There has been a good deal of discussion, not yet settled, as to whether the activation of the sperm cells is brought about by stimulatory substances secreted by the cells lining the tube of the epididymis, or merely by the process of maturing. Regardless of this question, it is at least certain that the epididymis is a favorable place for the sperm cells, for when experimenters have tied off its tube in two places, leaving sperm cells trapped between the ligatures, the cells have been found to remain active for two weeks or more.

Emerging from its coiling in the epididymis, the seminal canal becomes less tortuous and finally runs directly upward under the skin toward the groin, as shown in Fig. 30. This part of the system is called the seminal duct or *vas deferens*. The two ducts (one from each testis), pass over the front of the pelvic bones to enter the interior of the pelvis. Proceeding down the side and rear of the pelvis the two ducts approach each other under the urinary bladder. Before they unite each of them gives rise to a small saccular offshoot or branch, the *seminal vesicle*. These vesicles are club-shaped hollow structures, really side branches of the seminal duct. They are each about 10 centimeters (4 inches) long, but folded to half that length as they lie in place. They are glands of external secretion, producing in their cavities a clear gelatinous substance which becomes part of the seminal fluid. There is an old notion, hardly yet cleared out of the medical textbooks, that the seminal vesicles are reservoirs for sperm cells, but the fact is that sperm cells are not normally found in them.

The whole course of each seminal duct from epididymis to their junction, is about 30 centimeters (1 foot) long. As shown in Fig. 30, the two ducts unite just below the bladder and enter the urinary channel, the *urethra*, just after it makes its exit from the urinary bladder. The combined seminal and urinary channel then passes through the *prostate gland* and enters the penis.

The prostate gland. To most people the name of the prostate gland probably conveys no clear impression, but only a vague and slightly ominous suggestion of something one hardly speaks of unless it makes trouble. Years ago Mr. Henry Mencken, when a columnist on the *Baltimore Sun*, wrote an amusing article on the relative respectability of the human organs. The heart and lungs, he said, are perfectly respectable, the liver not quite, the spleen dubious and the kidneys definitely vulgar. In those days the prostate gland was so far below the standard of respectability that it could not even have been mentioned in the newspaper. Possibly the fact that its prosaic name, from the Greek *prostates* ("standing before" the urinary bladder), is often confused with the word "prostrate" adds to its flavor of indignity.

This unjustly disparaged organ is actually a gland of external secretion. It consists of 15 to 30 branched tubular glands, imbedded in connective tissue and muscle, forming a round mass 20 grams (2/3 ounce) in weight and almost completely surrounding the urethra just below the urinary bladder. The branching tubules of the prostate gland deliver a special secretion to the spermatic fluid, about which we know very little except that it is favorable to the activity and function of the sperm cells.

The various portions of the duct system and accessory glands are connected through the autonomic nerves so that all of them contribute to the seminal fluid when it is ejaculated at the climax of sexual excitement.

The prostate gland is one of those organs which carry on their useful functions in complete silence, never making known their presence or their action as long as they are in good working order; but when something goes wrong with one of them it suddenly becomes the focal point of the universe for the sufferer. Its bad reputation as a source of trouble in elderly men is due to the fact that for some obscure reason, probably of endocrine nature, the gland tends to enlarge in men past 50. Situated as it is around the urinary channel (see Fig. 30), and enveloped in a heavy capsule, which prevents it from swelling outward, any marked enlargement of the prostate inevitably blocks the outflow of urine, with serious consequences. The hope that prostatic enlargement may (when we know enough about it) be brought under control by treatment with hormones, lies temptingly before the investigators and may some day be realized.

Secondary sex characters. Before we can discuss the hormone of the testis we must take account of certain other matters that form part of the pattern of sex. Primarily a male animal differs from the female, or a man from a woman, because the one has in all his cells the chromosomes for maleness, the other the chromosomes for femaleness. One therefore develops testes, the other ovaries. The sex glands then begin, even in the early embryo, to call forth the secondary sex characters of their respective sexes. When the individual reaches the age of puberty these characters become prominent. In most mammals the males are larger, and possess heavier, rougher skeletons and stronger muscles. The shape of the pelvis, and to a lesser extent that of the skull and the other bones, is different in the two sexes. In humans the anatomy of the larynx is different and therefore the voice becomes either male or female. The distribution and growth of the hair are different. One sex has well developed mammary glands, the other only rudiments.

{ 227 }

In the various divisions of the animal kingdom there is a vast array of secondary sex differences. The tail of the peacock, the antlers of the stag, the beard and the smell of the billy goat, are evidences of what Nature can do in this way. The subject would fill a large book.[2] Perhaps the most familiar of all, the comb of Chanticleer, has been seized upon by the experimenters (as we shall see) and has been made to tell us, more than any other one sign, just how the hormones control the secondary sex characters.

THE HORMONE OF THE TESTIS

People have known since prehistoric times that castration of men and domestic animals suppresses the development of secondary sexual characters and causes atrophy of the accessory male sex organs, such as the seminal vesicles and prostate gland. The gelding of stallions, the castration of male calves to make steers, of cockerels to produce capons, and even of boys for the production of eunuchs, has long been practiced. If anyone asked how the testis can control the size and form of the skeleton, the distribution of hair, or the tone of the voice, the explanation was vaguely to the effect that some sort of "sympathy" existed between the parts of the body, with the implication that the nervous system is the connecting agent. In 1849, however, Arnold Adolph Berthold, a physician and zoologist of Göttingen, proved once for all that the influence of the testis is carried by something that travels in the circulating blood. Berthold's little contribution (it is only four pages long) belongs to the fundamental classics of endocrinology.[3] He tells us that on August 2, 1848, he castrated 6 cockerels, 2 to 3 months old. Their combs, wattles and spurs were not yet developed. From two of them the testes were completely removed. These became typical

[2] See many chapters of Edgar Allen, *Sex and Internal Secretions.*
[3] A. A. Berthold, "Transplantation der Hoden," *Archiv für Anatomie, Physiologie, und Wissenschaftliche Medizin,* 1849 (Appendix II, note 18).

capons, fat, docile, without cocks' combs, wattles and spurs, unable to crow. In the case of two others, Berthold removed both testes but put one of them back, dropping it among the intestines. Anatomical examination months later showed that the reimplanted testes had become attached to the intestines and had acquired a good blood supply, so that the testicular tissue flourished in its new site. Both these cockerels became typical cocks; they grew combs and wattles, crowed, fought their rivals, and, as Berthold delicately observes, "showed the customary attention to the hens." One of these was later opened surgically, the implanted testis was removed, and the comb and wattles cut off. The head furnishings did not regenerate, and the bird, now fully castrated, reverted to the status of a capon. The other two each had one testis removed, then Berthold exchanged the remaining testes, giving each bird the other's sex gland, which he implanted on the intestine. These also became typical cocks. This beautiful experiment showed that the testis by no means depends upon specific nerves to maintain its control of the secondary sex characters, but works through the blood.

A long story could be told of all the efforts that were made to follow up this discovery, and there would be many divagations to relate. There was, for example, the episode of Charles-Edward Brown-Séquard, a brilliant, restless Franco-Irish-American (1817-1894), who made two incursions into the field of the internal secretions. In 1856 he was the first to remove the adrenal glands from animals and to observe the fatal disorder thus produced, like an exaggerated Addison's disease. In 1889, when he was seventy-two years old, he began to dose himself with extracts of dogs' testicles. He was feeling the debility of age, and hoped to rejuvenate himself. Brown-Séquard had been a good scientist and it is almost incredible that he could have hoped to do critical experiments with himself as the only guinea pig, prejudiced by all his hopes and fears for his own health. He thought that after the injections he felt much stronger and more alert, and reported his experience with pathetic enthusiasm and intimate personal detail before the medical societies of Paris and in the journals. The medical profession was on the whole incredulous and Paris made a good deal of fun of him. We know now that the extract could have had little or none of the genuine testis hormone in it. It was made by putting mashed-up testes in water and filtering the mixture. We are now well aware also that when a man grows old he ages all over, not only in his testicles. Nevertheless the idea of administering gland extracts had its up-to-date appeal in those days of the earliest discoveries about the endocrine organs. German biochemists had recently isolated from animal testes a peculiar nitrogenous substance called "spermine." Various people leaped to the conclusion that this might be the active substance in Brown-Séquard's extracts. Spermine was therefore put on the market under the auspices of the chemist Poehl, and thus became (to the best of my knowledge) the first endocrine product to be commercialized. Thus Brown-Séquard's notoriety was probably responsible, more than anything else, for the exploitation of endocrine preparations in the drug trade ahead of scientific knowledge. Since then, barrelfuls of extracts and millions of tablets have been fed and injected into human patients, with uncritical optimism, before the chemists and physiologists could learn the facts. The benefits of endocrine research on the reproductive glands have almost been stifled by this exploitation. Even today the practicing physician finds it difficult to distinguish what is sound and practical amid the flood of well advertised endocrine drugs.

There have been premature efforts also to apply Berthold's experiment of transplantation of the testis to the rejuvenation of senile men. The most widely publicized of these was that of the Franco-Russian surgeon Serge Voronoff, who was busy from about 1912 to 1925 implanting monkey testes into human patients. The American journalists of those days could not refer to the testicles by name in the newspapers, and introduced the expression "monkey glands," which became a byword and finally a joke. Some of the patients reported hopeful results. The grafts may indeed have yielded a little of their hormone to the body before they disintegrated and disappeared. More often, no doubt, the benefit was entirely psychic. It is now clearly established that tissues from one species of mammal cannot grow in another species, and indeed it is practically impossible to secure a permanent transplant from one human to another. Grafting of the testis has therefore not been adopted as a sound procedure.

While we are on the subject of rejuvenation, we may as well mention the Steinach operation. Eugen Steinach of Vienna, a scientist of good reputation, came forward about 1920 with a proposal based on two premises. The first of these, which has even yet not been proved, was that the testis hormone is made entirely by the interstitial cells. (The question will be discussed more fully below.) The second premise has since been proved incorrect; it was that if the seminal duct is tied off, the seminiferous tubules will degenerate leaving more room for the interstitial cells, and these will increase in number and presumably make more hormone. Steinach believed that such a ligation of the seminal ducts of man (vasectomy) would restore vitality of body and of sexual function to elderly men. He brought forward apparently strong evidence from animal experiments to support the idea. The operation made a strong appeal to men who were yearning to regain their lost youth. It has been tried widely, but the medical profession remains unconvinced, and the scientific basis for it as outlined above has been disproved.

Meanwhile, through all this period of sensationalism and premature publicity, the slow implacable attack upon the problem by inconspicuous investigators has gone forward to notable success. The important thing in endocrine research is to find a test for whatever hormone one suspects to exist—some sharp-cut characteristic effect that is lost when the gland is removed and that can be restored by giving back the gland in implants or in extracts. At each step of our story in this book such a method has been the key to success. The vaginal smear test for the estrogens, the rabbit uterus test for progesterone, promptly made possible the purification of these substances. If the test is quick and cheap, the results will come that much faster. Berthold's experiment with the cockerels provided an ideal method of testing for a hormone of the male sex gland. Success came to those who followed his lead, putting aside premature efforts to work with slowly growing mammals or to rejuvenate old men. Between 1907 and 1927 two or three European investigators reported that they had extracts of the testis which induced growth of the head furnishings of cocks. None of these experiments, however, was fully convincing. Meanwhile the estrogenic hormones were isolated and discovered to be soluble in fat solvents. In 1927 a graduate student in biochemistry under Professor F. C. Koch at the University of Chicago, L. C. McGee (now a physician in Elkins, West Virginia) applied the new methods of extraction to the tissues of the bull's testis and promptly secured a relatively pure extract that was capable of producing rapid growth of capons' combs (Fig. 32). This lead was followed up by a group of workers in biochemistry and zoology in the University of Chicago, including McGee, F. C. Koch, C. R. Moore, L. V. Domm, and Mary Juhn, and by various workers abroad. The successive steps in the purification and identification of the male or *androgenic* hormones were much like those in the isolation of the estrogenic substances. In 1929, S. Loewe and S. E. Voss of Dorpat (Estonia) and also Casimir Funk and B. Harrow of New York found androgenic substances in human urine from males. The indefatigable Butenandt and his aides then

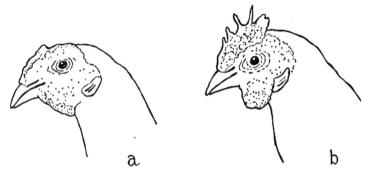

a   b

Fig. 82. The effect of testis hormone on the rooster's comb. *a*, untreated castrated cockerel. *b*, castrated cockerel after 11 days' treatment with testis extract. Drawn from photograph by Freud and coworkers.

isolated, completely purified and identified two of these compounds, called respectively androsterone and dehydroandrosterone, in 1931-1932. L. Ruzicka and his colleagues at Basel made them synthetically in 1934. The next year David, Dingemanse, Freud and Laqueur of Amsterdam succeeded in the difficult task of purifying the hormone from extracts of the testis itself, and found the substance called testosterone, which differs slightly in its chemical constitution from the androsterone found in urine. The groups of workers led by Butenandt and Ruzicka immediately synthesized this hormone as well. Now that such substances could be made in the test tube as well as in the testis, and (as we shall see) began to be found in other tissues also, the adjective "male" as applied to them gave place to the more apt word "androgenic," meaning "promoting masculinity"; the latter word defines the effects without implying any particular place of origin and can therefore be used of such substances when, for example, they turn up in female urine, in the cortex of the adrenal gland, or in a chemist's flask.

Chemistry of the androgenic hormones. These substances belong to the same family of chemical substances as the estrogenic hormones and progesterone. Testosterone has the following structure:

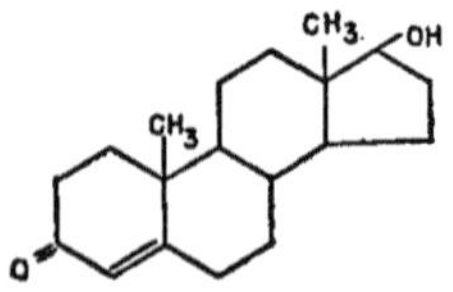

TESTOSTERONE

Androsterone, prepared from human male urine, has this formula:

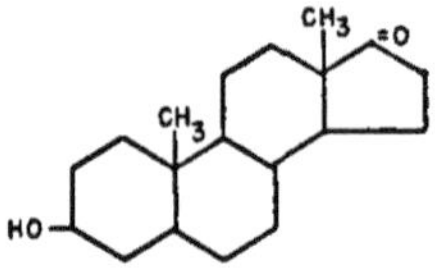

ANDROSTERONE

A statement of the relation of these particular sterols to simpler chemical substances will be found in Appendix I.

At the present time at least thirty-five substances of similar composition are known which have androgenic effects. Some of these have been found in the adrenal gland, or in the urine under special circumstances, but most of them are artificial products of the chemical laboratory. A list of them is given by Koch.[4] The male hormone which is actually at work in the body is probably testosterone itself or a closely similar substance. The reason a definite statement cannot be made on this point is that we are not sure that the chemical procedures necessary to extract and concentrate the hormone do not change it chemically. When testosterone is administered to a castrated animal, it is transformed in the body and is excreted by the kidney as androsterone. Since androsterone occurs in the urine of normal men, this is presumptive evi-

[4] Edgar Allen, *op. cit.*, Chapter XII, "The Biochemistry of Androgens," by F. C. Koch.

dence that testosterone, or something very much like it, is in circulation in the body.

Just what cells in the testis are responsible for producing the hormone is an interesting and debatable question. There are two possibilities, the interstitial cells on one hand, and the spermatogenic cells of the tubules on the other. Let us examine the evidence. In the first place the interstitial cells look like endocrine tissue, since they are large cells provided with a rich circulation of blood but obviously not producing an external secretion, for they are not arranged in channels and ducts. Years ago, moreover, Bouin and Ancel called attention to evidence favoring the interstitial cells as the source of the hormone. When the testes fail to descend and therefore do not form sperm cells, the cells lining the tubules are reduced to an inactive state (Plate XXIV, *B*). They assume a vague, nondescript form as if they were merely surviving without any function. The interstitial cells however remain in place, they look normal, and indeed have even been thought to increase in amount. Although the cryptorchid animal or man bearing such sex glands is sterile, he develops male secondary sexual characters and male sex psychology. With the spermatogenic cells seemingly inactive (as judged by appearances under the microscope) but the interstitial cells in good condition, it is difficult to avoid the assumption that the latter are making the sex hormone. When bits of the testis are grafted successfully into a castrated animal the same cellular state develops in the grafts, and the animal likewise develops male qualities. It was formerly thought that tying off the seminal ducts produced the same effect. The Steinach operation described above was based upon this whole set of considerations. It is now known, however, that blocking the ducts and thus damming up the semen does not stop spermatogenesis. It is also known that cryptorchid (undescended) testes with inactive sperm cell formation do not produce more hormone but considerably less. In short, the two functions of the testis, spermatogenesis and hormone production, run parallel to a certain extent. For this reason some of the soundest of the investigators are not willing to point to one or the other of the constituent tissues of the testis and say "here is the sole source of the hormone." The tubule cells, even if they are inactive in producing sperm cells, may for all we know still be taking part in making the hormone; or perhaps the two kinds of cells work together. If a time ever comes when the hormone can be recognized in the tissues in very small amounts, say by some sort of super-spectrograph or X-ray analysis, such as the modern magicians of the physics laboratories are using in their easier problems of test tube chemistry, then perhaps we can answer such questions as these.

There is no sharp division between the estrogens, the androgens, and the progestins. Several of these substances give both estrogenic and androgenic effects, and a few have been found even to affect the uterine lining like progesterone, though only in very large doses.

The androgens are usually assayed by testing their effect upon the growth of the comb of the capon. The League of Nations Committee on Standardization of Drugs set up in 1935 an international standard of potency, specified as the equivalent of 0.1 milligram of crystalline androsterone. This is approximately the daily dose required to give a measurable response in a capon's comb in 5 days.

The androgenic hormones are usually injected in oil solution. In recent years the method of implanting pellets under the skin has begun to be tried. The hormones are also effective when applied in suitable ointments. Growth of the cock's comb can be elicited by applying hormone-containing ointments directly to the comb itself (Appendix II, note 19).

Effects and medical use of the androgenic hormones. The

effects of the androgenic hormones, as C. R. Moore[5] neatly puts it, are measured by the difference between a castrated and a normal man. These substances substitute completely, in animal experiments, for the normal internal secretion of the testis. They counteract castrate atrophy of the seminal vesicles and prostate gland; when given to immature animals they stimulate precocious growth of the accessory sex organs, and they induce sex activity and mating in castrated and immature males. A castrated male, skillfully treated with potent hormones, will resemble a normal animal of his species in all respects except that he will be infertile because he is producing no germ cells.

These effects have been tested quite thoroughly in an experimental way upon human patients who lack the testis hormone. In these men and boys, as in laboratory animals, the injected hormones bring out all the known responses which the bodily tissues normally make to the natural hormones of the testis. It should be recalled here that the symptoms of deprivation of testis hormone may be due to defects in either one of two glands. The testes may themselves be missing or deficient, or the pituitary gland may be furnishing an inadequate supply of gonadotrophic hormone (see Chapter VI). In the latter case the testis will not be functional and the patient will show symptoms of testis hormone deficiency exactly as if the testis itself were the seat of the difficulty. At some time in the future, when endocrine treatment has reached a higher state of perfection, it may become possible to treat the pituitary cases with pituitary hormones, reserving the androgenic hormones for cases of deficiency of the testis itself. At present, however, the distinction is more or less academic. In either case the physician finds himself confronted with signs of male hormone deficiency, and the question of immediate importance is how

[5] C. R. Moore, "Physiology of the Testis," in *Glandular Physiology and Therapy*, 2d ed., Chicago, 1942.

far he can hope to help the patient by treatment with androgenic hormones.

I write about this subject of the medical use of the androgenic hormones with hesitation, because it is difficult on one hand to avoid raising false hopes without on the other hand underrating the progress that has been made and the future possibilities in this field. To make the problem clearer, let us consider a specific case.[6] Here is a boy in his middle 'teens who is not exhibiting the usual signs of sexual maturation. We know that if the deficiency persists he will grow to be a eunuchoid man. He will have underdeveloped genital organs, a delicate skin, the childhood type of hair distribution, a high-pitched voice; possibly also he will be overfat and he may suffer from muscular and circulatory difficulties causing easy fatigability and inability to do hard muscular work. Another important feature of his general immaturity will show itself in the long bones of the skeleton. The growth zones (epiphyseal junctions) will not close up at the usual time (18 to 20 years), but will go on growing, so that he will develop the long delicate bones of the eunuch and will thus display his defect in the entire configuration of his body. Although the deficiency has nothing to do with intelligence or fundamental character, he is very likely, as he grows up, to suffer from psychological damage due to a sense of defectiveness and of difference from other men.

If this boy could be treated as simply as a guinea pig is treated in the laboratory, we could control all these visible defects by administration of androgenic hormone. The treatment is fairly expensive and must be kept up by frequent injections or inunctions. Administration by pellets buried in the tissues may in time become practical, but is hardly yet ready for use. We are, moreover, not yet free from insecurity about possible danger from long continued administration of

[6] J. B. Hamilton, "Testicular Dysfunction," in *Glandular Physiology and Therapy*, 2d ed., Chicago, 1942.

the sex hormones through damage to the tissues of various organs. Finally, whatever improvement is gained by treatment will wear off fairly rapidly if the drug is discontinued. In one feature only, so far as known, the improvement is permanent; that is in the skeleton, for the growth of the long bones will be permanently stopped by closure of the epiphyseal junctions. Even this requires care and forethought, for if the treatment is begun too early, the epiphyses may be closed prematurely and growth stopped before the boy reaches full stature.

I have no doubt said enough to make it clear that the use of androgenic hormones to correct testicular hormone deficiency is very decidedly still in the stage of exploration, and that every case so treated is an experiment. This does not mean that the attempt should not be made by properly trained physicians and scientists who are in a position both to guard the welfare of the patient and to study the results with rigid scientific standards for future guidance. We have, of course, been considering an extreme case, one in which there is permanent total deficiency. The androgenic hormones are certainly going to be useful in many disturbances of less intensity and as collateral treatment. To make up for partial deficiencies, to tide over the acute deprivation which follows surgical removal of the testicles, to supplement the treatment in various cases of sexual disability and impotence—for such purposes the androgenic substances will no doubt be helpful in competent medical hands.

At any rate, the contemplation of these distressing deficiencies of the sex hormones must have impressed the reader anew with the thought that the normal processes of reproduction involve a remarkable series of linkages and coordinations within the body. It is indeed a subtle and potent chemistry by which the reproductive glands create the egg and the sperm cell and surround them with all that is needful of nourishment and protection to carry them through their critical journey from conception to birth. In this book we have traced the main outlines of these complex physiological patterns. We have seen the great advance of knowledge that has taken place in less than fifty years, by the efforts of faithful laborious men who worked in peace and quiet, giving their lives to the understanding and improvement of life. When I wrote these closing words, the guns were sounding all over the world. The scientists were dropping their instruments, or turning them perforce to the uses of death and wastage. But life goes on nevertheless, and the problems of life will be studied until the day comes we all dream of, when mankind may everywhere seek the truth undaunted by fear of war and oppression. In that day we shall gather the fruits of our labor. The childless wife, the ailing girl, the boy deprived of his birthright of sex by some failure of Nature's process, will call and not in vain for the help that science can bring them, and man shall understand at last the miracle of his birth.

APPENDICES

MORE ABOUT THE CHEMICAL STRUCTURE
OF THE SEX GLAND HORMONES

WHEN discussing the chemistry of the ovarian and testicular hormones, in Chapters IV, V, and IX, I tried first to make their general nature as clear as possible to readers who have not studied chemistry at all, and then I gave the structural formulas for the benefit of those who are familiar with organic chemistry. Most of my readers, however, probably belong to a middle category. They have studied the elements of chemistry in a college course that included a few weeks on the compounds of carbon, so that they can comprehend an organic formula, at least of the more familiar sort, especially if written out in full and its significant features are explained. They are on the other hand hardly prepared to grasp at once the full meaning of one of these complex and unfamiliar hormones or to perceive its relation to the simpler substances chiefly dealt with in college chemistry. For the guidance of such readers I propose to write the formulas of the more important sex gland hormones as clearly as possible and to explain their nature exactly as I had to have them explained to me when I found that nowadays even an anatomist must struggle with chemistry.[1] I assume that the reader recalls that the valence of carbon is 4, of hydrogen and the hydroxyl group (OH) is 1, and of oxygen 2:

$$-\overset{|}{\underset{|}{C}}-\qquad H-\qquad -OH\qquad O=$$

[1] In preparing this discussion, I have drawn freely upon the standard textbooks of organic and biological chemistry. See also *The Chemistry of Natural Products Related to Phenanthrene,* by L. F. Fieser, New York, 1936; *Sterols and Related Compounds,* by E. Friedmann, Cambridge, England, 1937; and *The Chemistry of the Sterids,* by Harry Sobotka, Baltimore, 1938.

and that he comprehends the general significance of simple formulas such as that of ethyl alcohol and benzene:

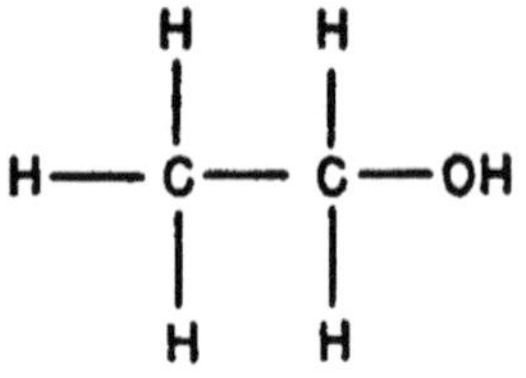

ETHYL ALCOHOL 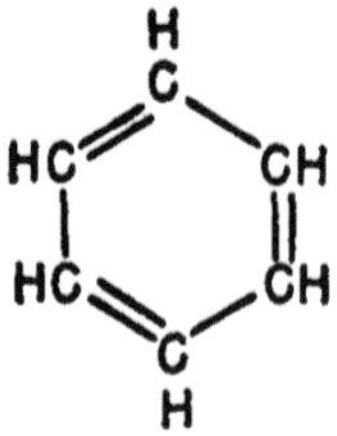   BENZENE

The basic hydrocarbon of the sex gland hormones. All the naturally occurring sex gland hormones can be understood in relation to certain basic hydrocarbons. One of these, called *estrane*, has the following structure:

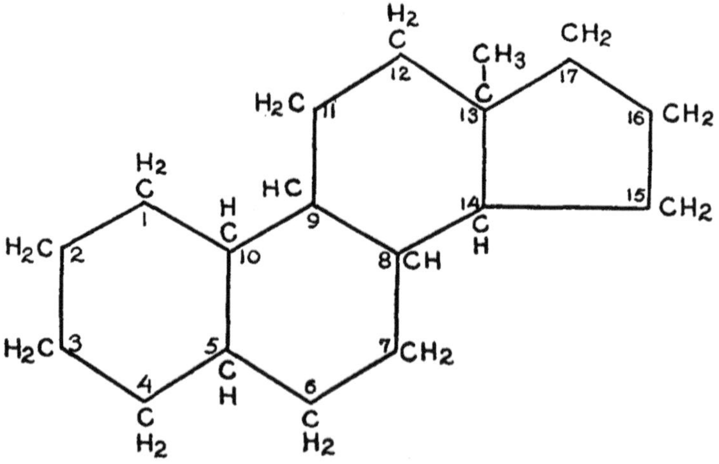

ESTRANE

$C_{18} H_{30}$

For the sake of convenience we number the carbon atoms arbitrarily, as shown in the diagram. We can save time and trouble henceforth by omitting the obvious CH's, as follows:

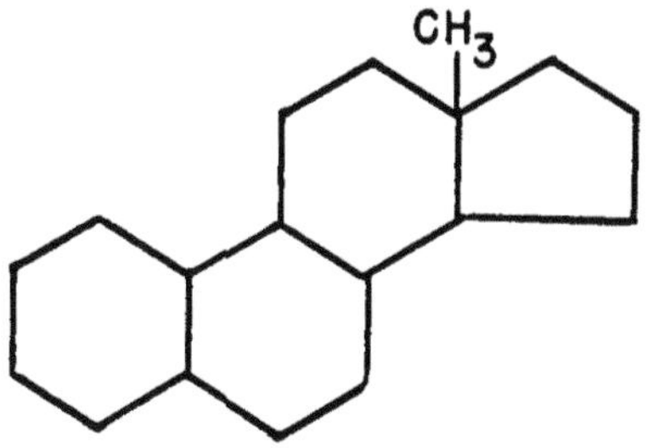

ESTRANE

This means the same as the complete formula above. Whenever, in formulas cited hereafter, a linkage point is written without indication of the elements, it means a carbon atom with enough hydrogen atoms to fill up its quota of 4 valences. Unsaturated carbon atoms will be indicated by double bonds.

The relation of estrane to simpler organic compounds. Those substances derived from estrane which are of interest to us as hormones belong to a group of compounds called sterols, which have already been characterized briefly in Chapter IV. The relation of the sterols to more familiar substances can be explained as follows:

Three benzene rings condensed in line form the substance *anthracene*:

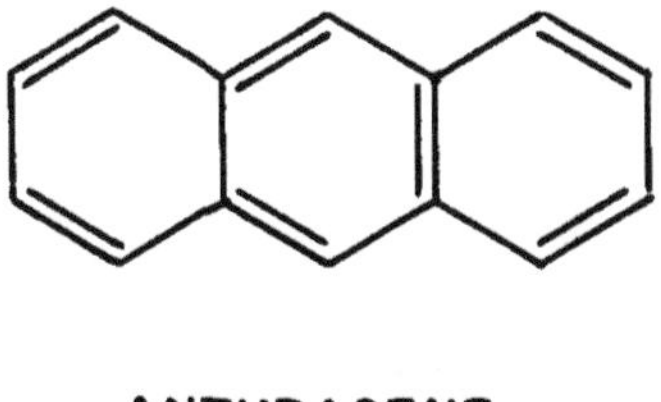

ANTHRACENE

In a slightly different arrangement, exactly the same atoms constitute *phenanthrene*:

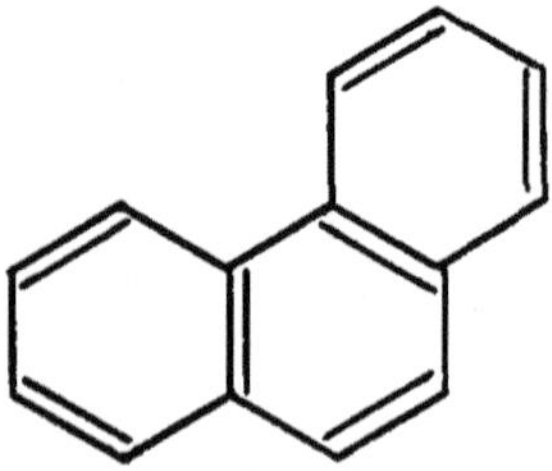

PHENANTHRENE

With one more ring, this time of 5 carbons instead of 6, we get cyclo-penteno-phenanthrene:

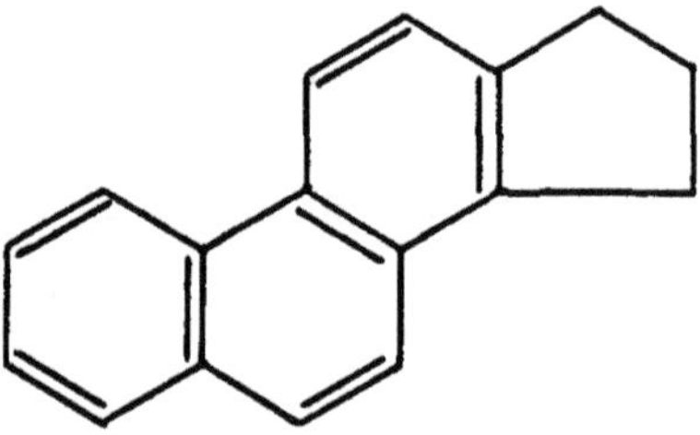

CYCLOPENTENOPHENANTHRENE

This structure of 4 rings, two upstairs and two downstairs, has a great significance to the chemist and physiologist, for it is the basis of many important substances in the body. When the rings are unsaturated at various linkages and provided with various side groups and side chains, an enormous series of compounds occurs, which are known as steroids. These include the sterols (one of which, *cholesterol*, is a widespread constituent of the animal body), the bile acids, and the sex gland hormones. The unravelling of their constitution has been one of the great feats of modern organic chemistry, and it is now being followed up by the rapid production of scores and hundreds of new synthetic compounds of similar constitution.

To proceed from cyclopentenophenanthrene to estrane, the hydrocarbon first mentioned above, let us imagine the former compound completely saturated with hydrogen and a methyl group (CH₃) attached to the carbon atom which we have numbered 13:

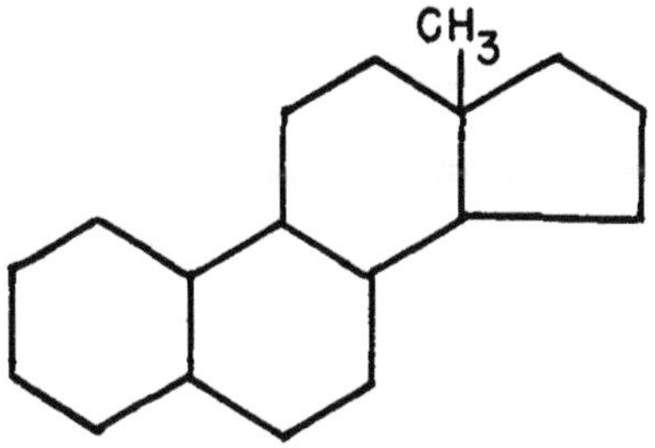

ESTRANE

If the first ring is unsaturated, we have *estratriene*:

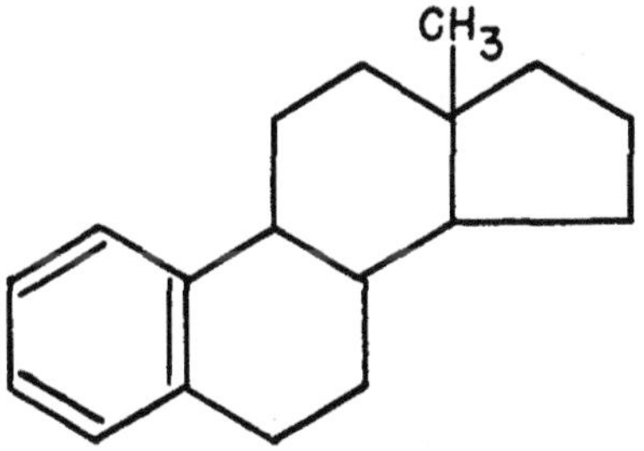

ESTRATRIENE

The estrogenic hormones. At last we have arrived at an actual hormone, for the best-known of all the estrogenic substances, *estrone*, is like estratriene except that it has a hydroxyl group at carbon 8, and an atom of oxygen at carbon 17; in other words it is 8-hydroxy, 17-keto estratriene (for clearness, the formula is written in full as well as in the usual simplified form):

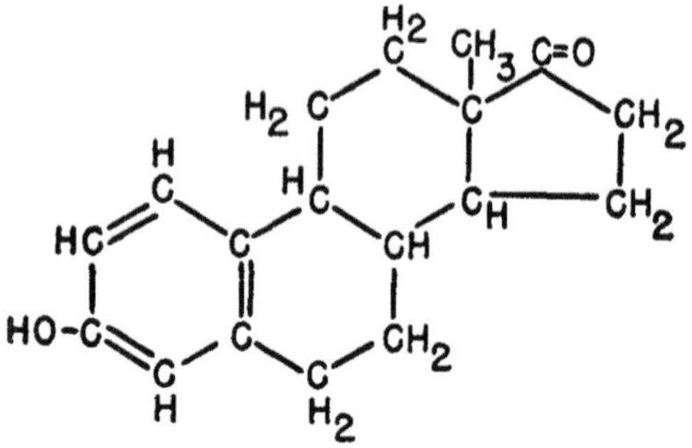

ESTRONE

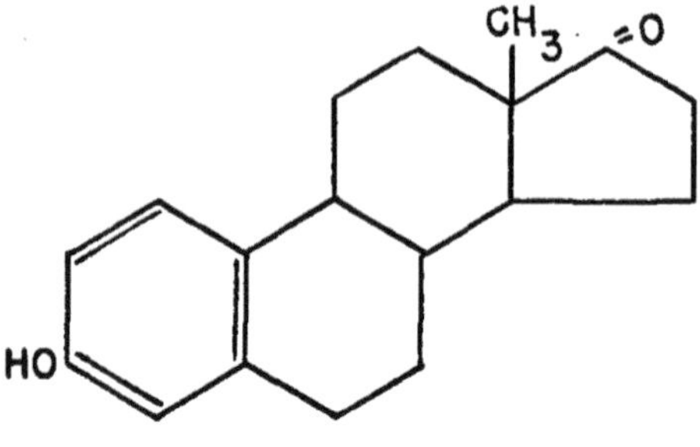

ESTRONE

As explained in Chapter IV, a large series of naturally occurring estrogens is known. All these are closely related to estrone. Among the representative estrogens are the following:

Estradiol, the hormone which probably exists in the greatest amount in the body, has a hydroxyl group at position 17 instead of the doubly-linked (ketonic) oxygen:

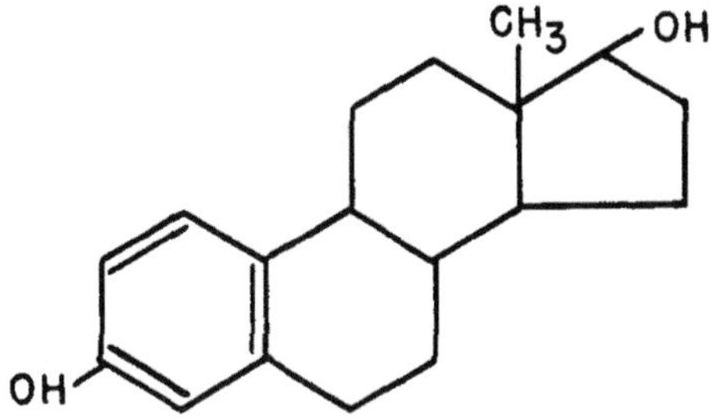

ESTRADIOL

Estriol, which occurs abundantly in the human placenta, is

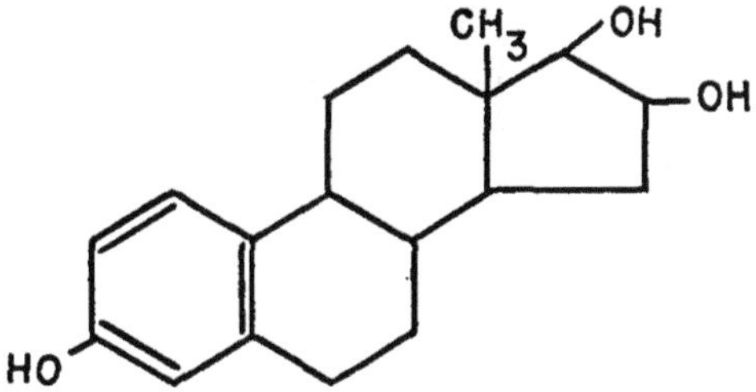

ESTRIOL

Equilenin, isolated from mare's urine, belongs to a group of estrogenic substances which have unsaturated carbon atoms in the second ring as well as the first:

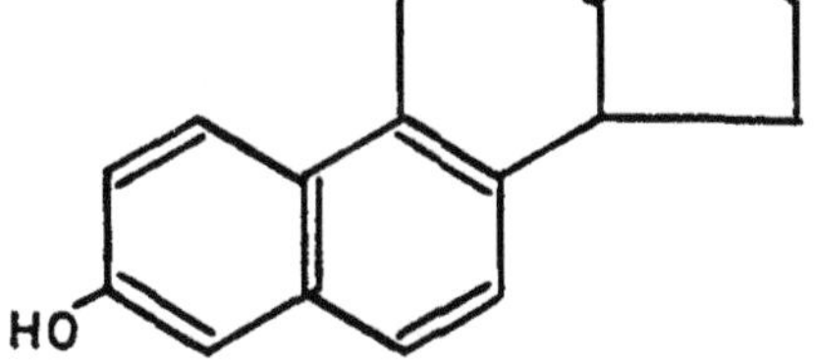

EQUILENIN

{ 249 }

With the foregoing explanation the reader, if he is still curious about these relationships and wishes to follow out such problems as the synthesis of estrone, is now in a position to study a more technical discussion of the chemistry of the estrogenic hormones such as E. A. Doisy gives in *Sex and Internal Secretions*.[2]

The androgenic hormones. These may be understood by reference to their basic hydrocarbon, *androstane,* which differs from estrane by having a second methyl group, attached to carbon 10:

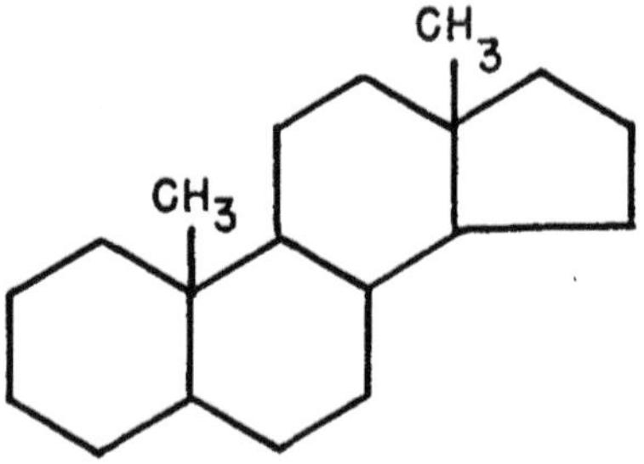

ANDROSTANE

Androsterone, referred to in Chapter IX, as the hormone of male type first isolated from human urine, is 3-hydroxy, 17-keto androstane:

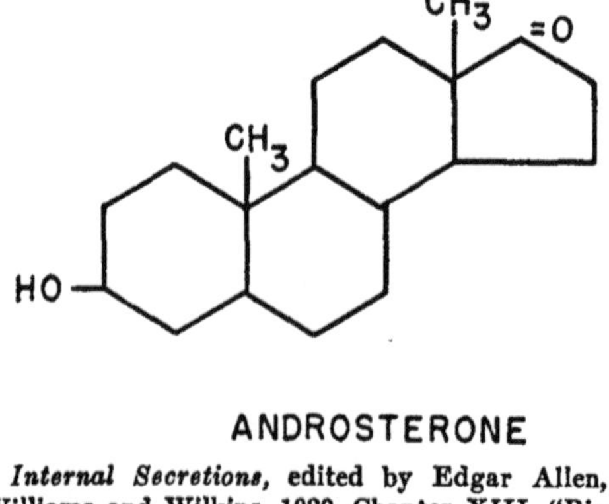

ANDROSTERONE

[2] *Sex and Internal Secretions,* edited by Edgar Allen, 2d edition, Baltimore, Williams and Wilkins, 1939. Chapter XIII, "Biochemistry of the Estrogenic Compounds," by E. A. Doisy.

Testosterone, the male hormone obtained directly from the testis, has a double bond in the first ring:

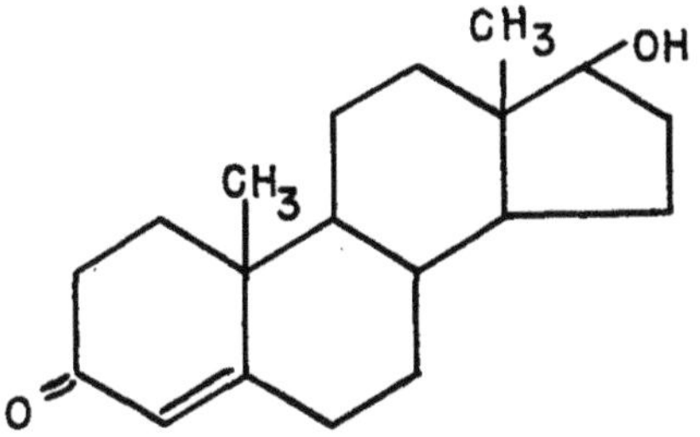

TESTOSTERONE

A large number of substances having androgenic potency, and the steps by which they have been prepared synthetically, are discussed by F. C. Koch in *Sex and Internal Secretions*.

Progesterone. The corpus luteum hormone can best be understood by comparison with its basic hydrocarbon, which is called *pregnane*. This has the two methyl groups already seen in androstane, and also an ethyl group ($CH_2 . CH_3$) as a side chain on carbon 17. (The two ethylic carbons are numbered 20 and 21, those of the methyl groups being 18 and 19.)

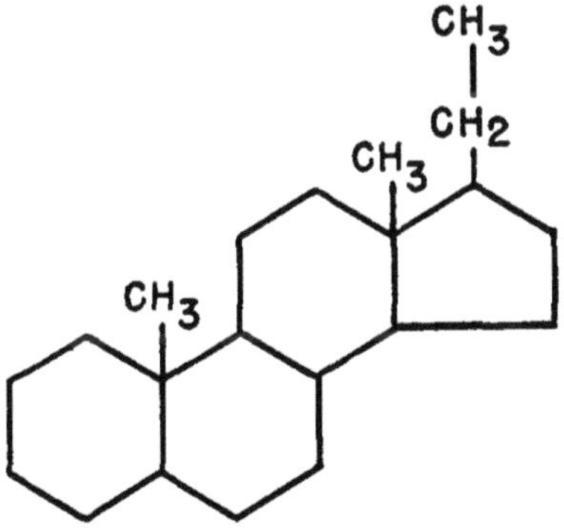

PREGNANE

Progesterone, as will be seen from the following formula, has a double bond in the first ring, and two atoms of oxygen at positions 3 and 20 respectively. It is therefore 3, 20 diketopregnene:

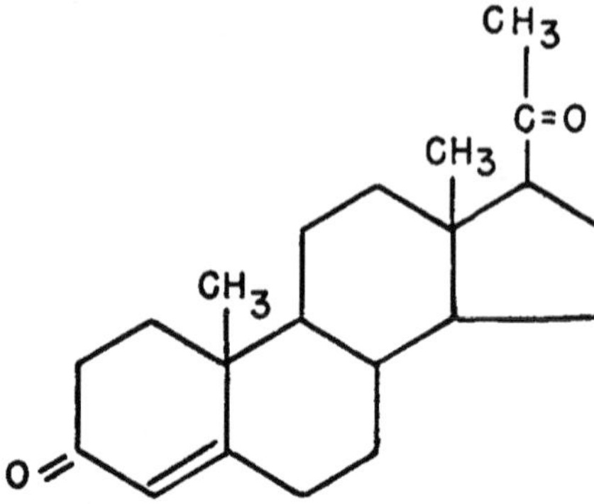

PROGESTERONE

Its excretion product in the human body, *pregnanediol*, can be derived from this formula by the addition of 6 atoms of hydrogen; that is, the 2 ketonic atoms are replaced by hydroxyl groups, the extra bonds at these points being satisfied by one hydrogen each, and the two unsaturated carbons are saturated:

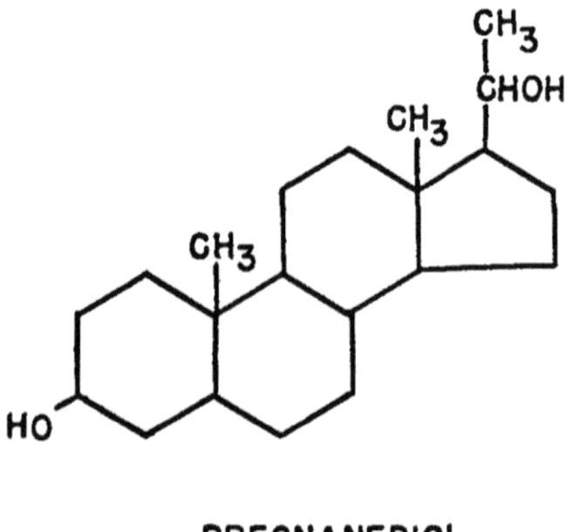

PREGNANEDIOL

The preparation of progesterone from pregnanediol and from stigmasterol (a natural vegetable sterol) and other chemical matters of interest concerning progesterone and its related substances are fully discussed by W. M. Allen in *Sex and Internal Secretions*.

As explained in Chapter V, when pregnanediol leaves the body in the urine, it does so in combination with glycuronic acid and with sodium. The resultant compound, sodium pregnanediol glycuronidate, has the following structure:

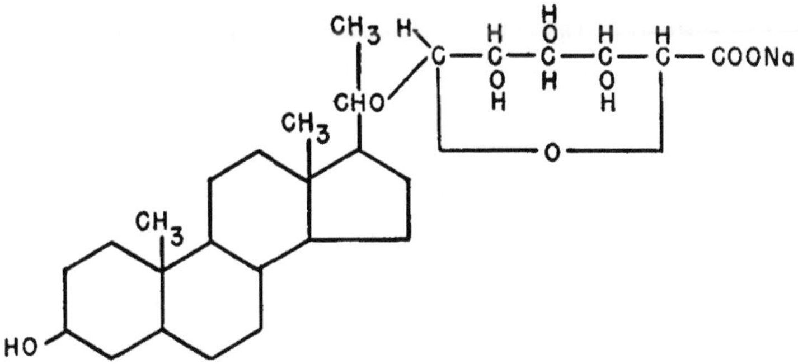

SODIUM PREGNANEDIOL GLYCURONIDATE

⊸§ APPENDIX II §⊷

Note 1 (Preface, page x, line 30). *Bibliographic refer-ences*. Full bibliographies covering practically the whole field of this book will be found in:

> *Sex and Internal Secretions*, edited by Edgar Allen, 2d edition, Baltimore, 1939.
>
> *Glandular Physiology and Therapy*, 2d edition, Chicago, 1942.
>
> *Biological Actions of Sex Hormones*, by Harold Bur-rows, Cambridge, England, 1945.
>
> *Endocrinologie de la Gestation*, by Robert Courrier, Paris, 1945.
>
> *Patterns of Mammalian Reproduction*, by S. A. Asdell, Ithaca, N.Y., 1946.

Note 2 (page 64, line 32). *The Atlantic palolo*. An in-teresting account of the swarming of an Atlantic species closely related to the oft-cited palolo of the Pacific Ocean, has recently been published by L. B. Clark and W. N. Hess, "Swarming of the Atlantic Palolo Worm," *Carnegie Institu-tion of Washington, Publication 524, Papers from the Tortugas Laboratories*, vol. xxxiii, 1943.

Note 3 (page 96, line 24). *Action of the sex-gland hor-mones on the embryonic sex organs*. Since the preparation of these lectures, there has been a considerable clarification of our knowledge of the action of the estrogens and of the male hormones (androgens) upon the growth of the accessory sex organs of the embryo and in particular upon their differ-entiation into male and female types.

Stating the fundamentals of the problem briefly, the original determination of the sex of an individual mammalian animal is made at the time of fertilization of the egg, by a mechanism which operates through the chromosomes of the egg- and sperm-nucleus. No visible difference between the sexes appears however until the embryo reaches an age (in the human species, about six weeks after fertilization) when the sex glands begin to show the characteristics of the ovary or testis respectively. The accessory sex organs, when they first develop, are capable of being directed toward the pattern of either sex. All the necessary rudimentary tissues for the development of both the male and the female sex organs are present in every embryo, regardless of the sex of the individual as determined by fertilization. About the 9th week of embryonic life (in humans) the sex-pattern of the accessory organs begins to be distinguishable, and after that the male organs (epididymis, vas deferens, prostate, seminal vesicles), and the female organs (uterus, vagina) begin to be recognizable to the embryologist, according to the sex of the individual.

Experiments on a number of species of laboratory animals, done at these early stages, reveal that before the definite characteristics of the accessory organs of the two sexes appear, and to a gradually diminishing degree thereafter, their differentiation may be modified and controlled by treatment with estrogens and androgens. For example, an embryonic animal which is genetically a male (as determined at the time of fertilization) may be made by suitable doses of estrogenic hormone to develop accessory sex organs of female type. Reversal in the opposite direction may be accomplished by subjecting a genetic female to treatment with androgenic hormone. The hormone administered by the experimenter thus overrides and contradicts the influences which normally control the sex-pattern of the accessory organs.

Such experiments were first done on the relatively accessible embryos of aquatic animals, especially the amphibians. The extensive literature which has grown up on this subject is well reviewed in *Sex and Internal Secretions*, Baltimore, 1939. Experimentation on mammals has been more difficult because of the inaccessibility of the embryos in the uterus. R. K. Burns, Carl Moore and others have cleverly taken advantage of the peculiar reproductive processes of the opossum. This creature, the only marsupial dwelling north of Mexico, gives birth to its young at the extraordinary age of 13 days after ovulation; that is to say, the embryos leave the uterus at that time and migrate to the brood pouch on the lower belly of the mother, where each embryo firmly grasps a nipple with its mouth and goes on growing and developing for many weeks. At the time of transfer to the pouch, the young are so embryonic that the accessory sex organs, including even the external genitals, are in the indifferent stage. The sex cannot be clearly determined by inspection until ten days after birth, when the scrotum of the male and the pouch of the female are definitely recognizable.

Once in the pouch, the young are accessible to the experimenter, who needs only to anesthetize the mother to get at them and inject them with micro-doses of the hormones. As in the salamanders and other lower vertebrates, so also in the opossum it has been possible to reverse the sex pattern of the accessory organs and to make genetic males acquire the structure of females, and *vice versa*. The gonads (testis or ovary) remain unchanged in their sex-affinity. In higher mammals, such as the rat, mouse, and guinea pig, in which the young are carried in the uterus until birth, very similar results have been obtained by the expedient of giving relatively enormous doses of the hormones to the mother during pregnancy. For detailed reviews of this subject the reader is referred to articles by R. K. Burns, "Hormones and the Growth of the Parts of the Urinogenital Apparatus in Mammalian Embryos," *Cold Spring Harbor Symposia on Quantitative Biology*, X, 1942; and C. R. Moore, "Gonad Hormones and Sex Differentiation," *American Naturalist*, 78, 1944.

These experiments teach us that the hormones of the kind produced by the adult sex glands can be used in the laboratory to induce differentiation of the originally indifferent accessory sex organs into structures typical of the male and female respectively. The experiments obviously do not tell us whether the substances that are produced in the embryonic gonads, acting under natural conditions to induce the development of the sex organs, are themselves hormones of the steroid estrogen-androgen types which are effective in adult physiology. For all we know at present, the embryonic hormones may be chemically like or unlike the known estrogens and androgens. This query has implications too large for further discussion here, leading us into the whole question of the nature of the embryonic inductor substances and their relation to growth-inducing and morphogenetic hormones. (For a brief elementary statement of the theory of embryonic inductors or "organizers," see G. W. Corner, *Ourselves Unborn*, New Haven, 1944, pages 103-106; for a detailed discussion, see Joseph Needham, *Biochemistry and Morphogenesis*, Cambridge, England, 1946.)

"*Paradoxical*" *effects*. It has frequently been observed, in experiments with the sex-gland hormones upon embryonic tissues, that higher dosages of the androgenic substances may induce female rudiments to accelerate development in the female direction, and there is evidence that estrogenic substances may sometimes act similarly upon male structures. The studies of Burns (see the article cited above) have shown that these so-called "paradoxical" effects occur in his embryonic opossums only when the dosage of the hormones is relatively high. By reducing the dose of the male hormone Burns was able to limit its stimulative effects to male structures only, all the female structures being unaffected. We may suppose therefore that the action of these hormones upon embryonic rudiments is fundamentally sex-specific, and that the "paradoxical" effects reported by various authors for a number of animals are the result, in some way, of excessive doses.

NOTE 4 (page 111, line 33). *Is the corpus luteum necessary for segmentation of the egg and for implantation of the embryo?* Nothing has turned up, since the first edition of this book was published, to cast any doubt upon the conclusion that the hormonal action of the corpus luteum is necessary for the survival of the early embryo in the uterus before its attachment, and for its implantation in the endometrium. A number of considerations make it necessary, however, to restate the matter somewhat differently. In a recent striking investigation M. N. Runner of the Jackson Memorial Laboratory at Bar Harbor (*Anatomical Record*, 98, 1947, p. 1) removed fertilized eggs of the guinea pig from the oviducts and placed them in the anterior chamber of the eye, where they could be observed with a microscope through the clear cornea. It has been known for some time that the fertilized mammalian egg can go on segmenting and reach the blastocyst stage outside the mother, in tissue-culture dishes. The motion pictures of W. H. Lewis (see page 56 and Plate XI) were made from eggs cultured in that way. The anterior chamber of the eye is well-known to be a favorable place for the growth of transplanted tissues, for the aqueous humor is an excellent physiological salt solution, and a good blood-supply is readily available for tissues that become attached to the iris (cf. Markee's grafts of endometrium, page 150). Runner found that when placed in the eye, the fertilized eggs of the guinea pig would continue dividing, would proceed to the blastocyst stage and become implanted, or at least begin to implant upon the iris. All these phenomena of embryonic growth were found, moreover, to occur even though the ovaries were removed or the eggs placed in the eye-chamber of another animal which had not ovulated and therefore had no corpora lutea. Indeed, the eyes of male mice proved to be as favorable for growth of embryos as those of females.

Some years ago J. S. Nicholas of Yale University operated upon rats in such a way that the fertilized eggs passed out of the oviduct into the abdominal cavity. Among a large number of animals thus prepared there were a few cases in which the embryos thus misplaced grew, attached themselves to the mesenteries or elsewhere on the peritoneal lining of the abdominal cavity, and lived out the full term of gestation.

The distinguished French biologist Robert Courrier in his *Endocrinologie de la Gestation* (Paris, 1945) reports a curious observation upon a rabbit in which, on the 19th day of pregnancy, one embryo was caused by surgical means to escape from the uterus into the abdominal cavity, where it attached itself to the peritoneum. Two other embryos were allowed to remain in the uterus. The ovaries were removed at this same operation. At the end of the usual term of gestation, the fetus in the abdominal cavity was alive, whereas those left in the uterus had died as a result of the loss of the ovaries. Evidently the corpora lutea are essential for the welfare of the embryos only if the latter are in the uterus.

We must assume from such experiments that the early mammalian embryo has a strong inherent vitality which will enable it to grow wherever it has the necessary warmth, a supply of oxygen and nutritive materials, and a generally suitable chemico-physical environment. Since, on the other hand, as experiments have proved, the early embryo cannot survive in the uterus except under the influence of pro-gestational changes induced by the corpus luteum, it follows that the uterus, when not so prepared, is actually an unfavorable place for the early embryo as compared with the anterior chamber of the eye, the peritoneal cavity, or even a tissue-culture slide.

This seemingly paradoxical situation is made intelligible by thinking of the evolutionary background of mammalian reproduction. It is characteristic of eggs and early embryos of lower animals that they are prepared to develop without shelter and nutriment from the mother. When the mammals evolved the phenomenon of utero-gestation, the chosen place of shelter, the uterus, was developed from part of the oviduct, a channel that had for its purpose the efficient transportation and discharge of the eggs, not their retention and maintenance. To fit it for gestational functions, the endocrine mechanism of the corpus luteum was evolved. In the light of this thought it is not surprising that the uterine chamber is actually a less favorable place for early embryos than, say, the anterior chamber of the eye, except when the hormones of the ovary act upon it and change it into a place of superior efficiency for its new function.

NOTE 5 (page 118, line 11). *Progestin by mouth.* The progesterone-like substance that can be administered orally is called pregneninolone (preg-nene-in-ol-one) or ethinyl testosterone. Its chemical structure is

PREGNENINOLONE

NOTE 6 (page 122, line 6). *Excretion products of progesterone.* The statement in the text is not quite correct. Marker, Wittle, and Lawson have found pregnanediol in the urine of pregnant cows and mares, in concentrations not greatly different from those in human pregnancy urine. They found, strangely, that the urine of bulls contains about twice as much pregnanediol as human pregnancy urine. It unfortunately remains true that the end-products of progesterone metabolism have not been identified in the common laboratory animals.

NOTE 7 (page 126, line 2). *Is the corpus luteum necessary throughout pregnancy?* Experiments on the Rhesus monkey by C. G. Hartman and G. W. Corner, which were under way when the first edition of this book was published, have been completed (*Anatomical Record*, vol. 98, August 1947). They prove that the corpus luteum of pregnancy, and probably the whole of the ovarian tissue, can in this species be removed as early as the 25th day of pregnancy without disturbing gestation.

NOTE 8 (page 132, line 11). *Clinical use of progesterone and pregneninolone.* Five years after these pages on the practical use of progesterone were first written, they may be reprinted with little or no modification. The same hopes, the same successes, the same cautions, still stand. Some progress has been made in selecting cases suitable for progesterone therapy, thanks to the use of pregnanediol excretion as a test of the need for progesterone. In the large university hospitals, where there are laboratories in the women's clinics equipped for the assay, only those cases of habitual abortion and of menorrhagia that show a low excretion of pregnanediol are treated with progesterone or pregneninolone. In these selected cases, naturally, the percentage of favorable results is higher than when all cases of a given disease are treated on a hit-or-miss basis.

Sterility, when there is similar evidence that a low progesterone level is involved, is to be added to the list of pathological conditions in which progestin therapy is now being tried.

NOTE 9 (page 135, line 24). *Menstruation in lower primates.* Evidence has accumulated that something like menstruation, in an elementary form at least, occurs in the New World monkeys. In several species of howler, spider and capuchin monkeys there is periodic shedding of small amounts of blood into the tissue of the lining of the uterus. A few red blood cells escape into the genital canal and can be detected in washings made by injecting salt solution into the vagina, removing it again and examining it under the microscope. There is not enough blood lost to show externally.

A process apparently resembling menstruation in the elephant shrew of South Africa has recently been described by Van der Horst and Gillman. The species in question belongs to a family of animals which has been assigned by some naturalists to insectivores and by others to the primates.

In summary, it begins to appear that menstruation is not sharply limited to the higher primates, but that on the contrary it exists in a rudimentary form in other families of primates and primate-like animals. (See S. A. Asdell, *Patterns of Mammalian Reproduction*, Ithaca, N.Y., 1946.)

NOTE 10 (page 137, line 26). *Menstrual cycles in infra-human primates*. Thanks largely to the work of Zuckerman and Gillman, the cycles of two species of baboon have now been studied in numbers large enough to warrant comparison with other primates. The cycles are longer than those of the human species and the Rhesus monkey, averaging about 33 to 36 days, modal length, in various groups of animals. (The subject of menstrual cycles in primates is thoroughly reviewed in S. A. Asdell, *Patterns of Mammalian Reproduction*, Ithaca, N.Y., 1946.)

NOTE 11 (page 141, line 33, and page 201, footnote). *The gonadotrophic substances of the pituitary gland and the placenta*. When this book was first written, it was thought best in the interest of clarity not to refer in detail to the moot question of the existence of two or more gonadotrophic substances. The problem is still not settled, but it has become somewhat better defined, and for the sake of readers who wish to proceed from the very general account given here, to the more technical literature, the following outline is now supplied.

The prime fact is that the pituitary gland, the placentas of certain species, the urine of pregnant females of certain species, and the blood serum of the mare, all contain hormonal substances that stimulate the growth and differentiation of the ovaries and testes of animals into which they are injected. The precise nature of the effects of these hormones differs considerably, however, according to the particular source of the hormone and also according to the species of animal receiving the injections and the dosage. Under these varying circumstances, the constituent tissues of the ovary, for example, respond differently. Sometimes the follicles merely grow larger or even become cystic or atretic; in other experiments they are caused to form corpora lutea. In the testis likewise, there may be stimulation on one hand of the spermatogenic tubules, on the other hand of the interstitial cells (see Chapter IX). Workers using the ovaries of various species as test objects have found first that as their efforts to purify the gonadotrophic substances advanced, they seemed more and more clearly able to achieve at least partial separation of the two effects just mentioned; that is to say some of the partially purified preparations tended to produce only follicle stimulation, others only to cause luteinization. Separation of the two effects, perhaps even better than can be attained as yet in the chemist's flasks, is observed as a result of biological processes. To mention one example, urine of women after the menopause, or after removal of the ovaries, contains a substance presumably produced by the pituitary gland that has almost pure follicle-stimulating properties. The hypothesis has therefore sprung up that the pituitary gland produces two distinct hormones, usually denoted respectively by the initial letters FSH for "follicle-stimulating hormone," and LH for "luteinizing hormone." There is ample evidence that the gonadotrophic substances are of protein nature, and this is sufficient explanation of the fact that as yet, in spite of immense effort, no one has obtained preparations which solely give one or the other effect upon test-animals. The question therefore remains unsettled whether there are two hormones, chemically separable, or one which acts differently under different circumstances; but the hypothesis that there are two has been generally accepted as a working basis for further chemical research and for speculation about the part played by the pituitary gland in the menstrual cycle and in pregnancy.

In the human species, the urine acquires during pregnancy a gonadotrophic potency which appears to depend upon a mixture of properties resembling those of FSH and LH. The same is true for limited periods in the pregnancy of chimpanzees and Rhesus monkeys. Whether the substances producing these effects ("Prolan A" and "Prolan B" respectively) are chemically and biologically identical with FSH and LH, has been debated, and this question, like many others relating to the gonadotrophic hormones, awaits the final purification and identification of the substances. The prolan-complex is undoubtedly produced by the tissues of the placenta.

The blood serum of pregnant mares contains a gonadotrophic substance, believed to be produced by the placenta, which because of some chemical peculiarity does not get into the urine. This substance, known as equine gonadotrophin or PMS (for pregnant mare serum), acts upon test-animals as if it were a mixture of an FSH-like hormone with a small proportion of LH-like material. The serum of pregnant mares has been much used as a source of gonadotrophic hormones by experimenters and drug manufacturers.

A good review of this subject, bringing it up to 1945, will be found in Burrows, *Biological Actions of the Sex Hormones*, 1945.

NOTE 12 (page 153, line 9). *The coiled arteries.* New questions about the theory of menstruation, and particularly about the role of the coiled arteries, that have arisen since the first writing of this book, are fully discussed in Note 13. At this point, however, it will be well to modify the assumption that the coiling of the arteries is essential to menstruation. In a pending article, a fellow-worker of the author, Dr. Irwin H. Kaiser, shows that the corresponding arteries in certain New World monkeys that undergo at least a rudimentary type of menstruation, are not coiled. Some, moreover, of the current hypotheses about the structure and function of the arteries in women and in Rhesus monkeys, to be mentioned in Note 13, do not depend upon the coiling.

The reader should therefore substitute for the statement in our text that "menstruation is primarily an affair of the coiled arteries" the more cautious and less specific thought that menstruation is primarily dependent upon special peculiarities of the arterial circulation of the endometrium, meanwhile keeping his mind open until this fascinating problem is further elucidated.

NOTE 13 (page 170, line 27). *Current thought about the mechanism of menstruation.* The hypotheses set forth in the original edition of this book may still be read profitably, except that the reader should substitute the term "endometrial arteries" for "coiled arteries" because (as explained in Note 12) there is evidence that the coiling is not *per se* essential to the menstrual process. The whole subject of the cause of menstruation has been actively revived in 1946-1947, largely as the result of studies made in Copenhagen by a group of anatomists and pathologists who did not let the German occupation stop their research.

Most of the thinking still involves the idea that the periodic menstrual flow results from some peculiarity or other of the endometrial arteries which makes them dependent upon hormonal support. As already mentioned, a view now somewhat in disfavor held that it is the coiling of these arteries which renders them sensitive to fluctuation of the ovarian hormones. It was conjectured that as the endometrium grows thicker in each cycle under the influence of estrogen, the coiling becomes more intense until the flow of arterial blood is impeded, the capillary circulation is impaired, ischaemia of the endometrium results and is followed by menstrual breakdown. As a variant of this idea, it has been thought that a drop in estrogen occurring toward the end of the cycle causes involution of the endometrium, with a reduction of its thickness and consequent tighter coiling of the arteries. This was supposed to cause damage to the tissues and consequently to bring on menstrual bleeding.

Another totally different supposition is now put forward by the Danish investigators Schlegel, Dalgaard, and Okkels (see, for instance, J. V. Schlegel, "Arteriovenous Anastomoses in the Endometrium in Man," *Acta Anatomica*, vol. 1, 1945-46). These workers, using very careful methods of injecting the uterine blood vessels, have shown almost beyond any doubt that in the human endometrium there are frequent direct connections (anastomoses) between the terminal arterioles and the venous spaces from which the uterine veins take origin. Some of the blood flowing through the lining of the uterus follows the pathway usual in other organs and tissues, through the capillary blood vessels, thus serving the ordinary metabolic functions of the blood. Some of the blood, however (according to these investigators) passes directly through a shunt, as it were, into the veins. Schlegel offers a theory of menstruation based on this finding. He conjectures that as the endometrium grows thicker in each cycle under the influence of estrogen, the number of short circuits between the arterial and venous systems increases. The increasing proportion of blood thus shunted must be compensated for by an increased flow through the capillaries also. Such a flow, it is well known, is effected by the estrogenic hormone. The time comes, however, it is thought, when the estrogenic stimulus is not able to produce further capillary flow whereas the shunts still divert much of the blood. The tissues nourished by capillary blood suffer injury and menstruation is thus initiated.

A variation of this hypothesis, suggested by Professor Okkels, involves also a vasoconstrictor substance (cf. Hypothesis 3, page 171).

American workers of the Bartelmez-Daron school, who have not observed arteriovenous anastomoses in their material, obtained chiefly from monkeys and prepared with great care though by methods differing from those of the Danish investigators, are naturally doubtful of hypotheses that depend upon the arteriovenous shunts.

Other possible mechanisms that have been hinted at, but not as yet supported by thorough anatomical demonstration, depend upon supposed peculiarities of the walls of the endometrial arteries, which are indeed somewhat different in microscopic structure from those of other organs. It is thought, at least vaguely, that their general structure requires in some way the support of estrogenic hormones, or that there are special points or regions on the arteries which are sensitive to hormone fluctuations and thus serve as sphincters to shut off arterial flow.

Enough has been said to show that while the relation of the uterine arteries to the menstrual process is still unsolved, the question is being actively studied and we may hope for better knowledge by the next time this book requires revision.

NOTE 14 (page 172, line 17). *Toxin theory of menstruation.* O. W. Smith and George Van S. Smith have modified their conjectures about the cause of the menstrual breakdown of the endometrium. As explained in an article in *Clinical Endocrinology*, 1946, vol. 6, they suggest that catabolic changes of the endometrium resulting from reduction of estrogen at the end of the cycle cause the formation of a toxic substance which damages the finer blood vessels and thus brings on the menstrual necrosis and hemorrhage.

Note 15 (page 185, line 34). *Amount of progesterone secreted daily in the human.* It is now known that by no means half the progesterone that gets into the blood is excreted in the urine as pregnanediol. If a measured amount of progesterone is administered by injection, only about 10 to 15% of it appears as pregnanediol. On the basis of such results G. Van S. Smith and O. W. Smith, Seeger-Jones and Te Linde, and others now estimate that the corpus luteum secretes about 50 milligrams of progesterone per day at the peak of its cyclic activity.

Note 16 (page 194, line 33). *Amount of estrogen produced daily in the human.* G. Van S. Smith, O. W. Smith, and Sara Schiller, *American Journal of Obstetrics and Gynecology,* vol. 44, pp. 605-615, 1943, published an estimate based admittedly on a number of unproved assumptions concerning the metabolism of the estrogens. Their result, 0.08 to 0.70 milligrams, averaging 0.33 mg., is not far from that reached by my totally different method of estimation; my figure, expressed as estrone, is equivalent to 0.30 milligrams.

Note 17 (page 211, line 12). *The isolation and identification of prolactin.* Two months after these Vanuxem lectures were delivered, White, Bonsnes, and Long of Yale University announced success in the isolation from beef pituitary glands of a crystalline substance of high lactogenic activity. The hormone is a protein of high molecular weight (32,000 or more). Readers with a knowledge of biochemistry will be interested in their account of their own work and that of Lyons and other investigators upon which their efforts were partly based. (*Journal of Biological Chemistry*, vol. 143, 1942, pp. 447-464).

Note 18 (page 228, last line of footnote). *Berthold's article.* A translation of the original paper into English, by

D. P. Quiring, was printed in the *Bulletin of the History of Medicine*, Baltimore, vol. xvi, 1944, pp. 399-401.

NOTE 19 (page 236, line 33). *Oral androgen.* Androgenic hormones that can be taken by mouth in tablet form have been prepared by chemical synthesis and are on the market.

INDEX

⊰ INDEX ⊱

9 780691 627724